D0933529

Consciousness Demystified

Consciousness Demystified

Todd E. Feinberg and Jon M. Mallatt

The MIT Press
Cambridge, Massachusetts
London, England

This book was set in ITC Stone Sans Std and ITC Stone Serif Std by Toppan Best-set Premedia Limited. Printed and bound in the United States of America.

Library of Congress Cataloging-in-Publication Data

Names: Feinberg, Todd E., author. | Mallatt, Jon M., author.
Title: Consciousness Demystified / Todd E. Feinberg and Jon M. Mallatt.
Description: Cambridge, MA : MIT Press, [2018] | Includes bibliographical references and index.
Identifiers: LCCN 2018007481 | ISBN 9780262038812 (hardcover : alk. paper)
Subjects: LCSH: Consciousness—History. | Evolution.
Classification: LCC BF311 .F375 2018 | DDC 153—dc23 LC record available at https://lccn.loc.gov/2018007481

10 9 8 7 6 5 4 3 2 1

Contents

Preface

Most scientists who study consciousness agree that at some point in the history of life on Earth, animals that have consciousness evolved from simpler animals that did not have it. Yet despite this generally accepted fact, many scholars from diverse disciplines have argued that there is something fundamentally different about the nature of consciousness. Unlike third-person objective reality, there appear to be seemingly unbridgeable "gaps" between the material neurons of the brain and seemingly immaterial experience.

Whereas our previous book, *The Ancient Origins of Consciousness* (MIT Press, 2016), was more about exploring which animals are conscious and when consciousness evolved, this one centers on the nature of consciousness and these apparent explanatory gaps between the brain as an *objectively observed* biological organ and *subjective* experience.

Here we try to explain how, despite the controversies that continue to swirl around the mystery of consciousness, the subjective, first-person, experiencing aspects of consciousness are created by natural brain processes, how they evolved in natural ways, and how the aspects of subjective experience that *appear* to be unexplainable by known brain mechanisms can in fact be naturally explained. We call our theory *neurobiological naturalism*, and it is an outgrowth and elaboration of John Searle's theory of biological naturalism. We attempt to explain that while subjective experience is *unique in nature*, there is no need to invoke any new, unknown forces, either physical or nonphysical, to account for its creation. We also link consciousness strongly to life, as a characteristic that evolved in living animal bodies.

Over the last six years or so, we have written a number of scientific publications on the nature and evolution of consciousness. Here we summarize and build on the major findings of our previous works in a more condensed format, using less technical detail, with more emphasis on the main point of how consciousness can be explained. This approach should make it easier for the reader to delve into and understand the nature of consciousness. To help with this goal, we have also included a glossary of the important terms and concepts at the end of the text.

This book adds to and updates our ideas with more findings from the scientific literature published in the several years since we researched our last book. For example, scientists today better understand the nervous circuits responsible for affects (emotional feelings). The neural pathways for the other main aspect of consciousness—building and experiencing mapped mental images of the world—were worked out at least thirty years ago, so it is exciting that the knowledge of affects is catching up with the knowledge of images.

Our primary goal in writing this book is simple: to demystify and naturalize consciousness. We aim to do this by placing its most perplexing philosophical features squarely among the properties of life and explaining when and why subjectivity evolved and how experience is created.

Acknowledgments

We thank our artist at the Icahn School of Medicine at Mount Sinai, Jill Gregory, for her sterling artwork in this book. Trained as a medical illustrator, she has become a fine animal artist, and her color work is superb.

Todd Feinberg thanks his wonderful wife, Marlene, who—as always—is his best friend and ally, and his daughter Rachel, son Josh, and grandson Jake, all of whom continue to inspire him. Jon Mallatt thanks his devoted and ever-young wife, Marisa de los Santos, for her constant support and help, and his daughter Justine for her great love of life.

At the publisher, we especially thank Phil Laughlin, who brought this, our second, book on the subject of consciousness to the MIT Press and expertly guided it through the review process. We also thank four anonymous reviewers, whose suggestions helped us to recast the book from a somewhat different angle and to clarify its thesis and organization. We are most fortunate to have had Judith Feldmann as the production editor and William G. Henry as the copy editor.

1 What Makes Consciousness "Mysterious"?

When we compare consciousness to the rest of biology, there appears to be something different or unexplained about it. But while we agree that consciousness indeed differs in many ways from the rest of biology—and we will explain what these differences are—we should also point out that biological processes themselves have many unique aspects when compared to the rest of the nonbiological natural world.

The evolutionary biologist Ernst Mayr noted in his book *What Makes Biology Unique?* that life has properties and functions that are in fact *unique* to biology among the sciences: reproduction, coded information, metabolism, adaptation to the environment, and so on.[1] Yet despite these "special" and unique features of life, the consensus among biologists is that all life processes are naturally explainable.

As Mayr explains, for centuries scholars found life to be so mysterious that many believed in a fundamental "life force" that both animated matter into life and built developing embryos into complete human bodies. But in the twentieth century, great leaps in understanding the biochemistry of life eliminated any need for belief in a life force.

However, while the scientific basis of life is no longer a philosophical or scientific mystery, in the case of consciousness—more specifically in the case of *subjective experience*—the discontinuity between the brain as an objectively observable biological organ, and the subjective experiences that the brain produces, appears to many scientists and philosophers to be mysteriously unbridgeable. For many scholars, an unexplainable divide always appears to exist between the objective brain and the first-person, subjective experience of consciousness. And while biology in general and nonconscious brain functions such as neural reflexes can be explained completely by the known mechanisms of physics, chemistry, and biology, consciousness displays perplexing features that defy conventional scientific explanation. That is, whenever one tries to explain the relationship between the neurons of the brain and subjective experience, something is always scientifically "left out" of the equation. The philosopher Joseph Levine called this the "explanatory gap" between the physical brain and the qualitative aspects of subjective experience that it creates.[2]

In this book, we are primarily interested in explaining the "gap" between the brain and the most basic forms of subjective sensory experience, what the philosopher Thomas Nagel called "something it is like to be" in his famous paper "What Is It Like to Be a Bat?" (1974):

> But no matter how the form may vary, the fact that an organism has conscious experience at all means, basically, that there is something it is like to be that organism. ... Fundamentally an organism has conscious mental states if and only if there is something that it is like to *be* that organism—something it is like *for* the organism. We may call this the subjective character of experience.[3]

This form of basic experience has various names, including *sensory consciousness*, *phenomenal consciousness*, *primary consciousness*, and *perceptual consciousness*.[4] We will use the terms "primary consciousness" and "sensory consciousness" in this book. Primary consciousness requires no self-reflection, so it is not the same as the more evolved "higher" forms of consciousness such as self-awareness or thinking about one's thoughts. Rather, primary consciousness is simply the capacity to have any experience at all. As the philosopher Antti Revonsuo described it:

> The mere occurrence or presence of any experience is the necessary and minimally sufficient condition for phenomenal consciousness. For any entity to possess primary phenomenal consciousness only requires that there are at least *some* patterns—any patterns at all—of subjective experience *present-for-it*. It is purely about the *having* of *any* sorts of patterns of subjective experience, whether simple or complex, faint or vivid, meaningful or meaningless, fleeting or lingering.[5]

That said, when we try to explain even this most basic and elemental form of consciousness, there still appears to be a mysterious divide between the brain and subjective experience. Our main goal here is to explain and "naturalize" that gap.

Searle's Biological Naturalism

It goes without saying that innumerable philosophers and scientists have tried to explain the divide between the brain and subjective experience. However, our approach follows the example of the philosopher John Searle, who has a theory that he aptly calls *biological naturalism*.[6] Searle argues that no conscious feeling states of any kind can be equated with or reduced to the brain. "Reduced to the brain" means fully explaining consciousness by the functions of the parts of the brain. This is impossible, says Searle, because a third-person (objective) observation of the brain inevitably leaves out this first-person (subjective) experience that we want to explain in the first place:

> Biological brains have a remarkable biological capacity to produce experiences, and these experiences only exist when they are felt by some human or animal agent. You can't reduce these first-person subjective experiences to third-person phenomena for the same reason that you can't reduce third-person phenomena to subjective experiences. You can neither reduce the neuron firings to the feelings nor the feelings to the neuron firings, because in each case you would leave out the objectivity or subjectivity that is in question.[7]

Nonetheless, according to Searle, despite this permanent gap between first-person experience and third-person observation, mental phenomena are entirely natural and are caused by neurophysiological processes in the brain.

Searle's assertion that consciousness can be "naturalized" means to him that mental events and processes are as much a part of our biological natural history as digestion, mitosis, meiosis, enzyme secretion, or for that matter any other biological process, and therefore no undiscovered fundamental principle of science is required for a complete scientific and philosophical account of consciousness. But while we agree with his basic premise, we feel that Searle does not pay enough attention to the special status of consciousness within the biological sciences. That is, he does not consider what unique factors of the conscious brain could account for something as singular as subjectivity. Surely without that sort of explanation, consciousness will remain a mystery to philosophers and neuroscientists alike.

Neurobiological Naturalism: A New Synthesis

Our theory, which says that consciousness is fully natural but requires explanations that are uniquely different from explanations applied to the rest of biology, is called *neurobiological naturalism.*[8] It attempts to explain the uniqueness of subjectivity in the natural universe, but without invoking any new scientific laws or principles. Thus, in certain respects, our approach takes after that of Mayr, who accepted the uniqueness of life processes but did so in a thoroughly naturalistic context. Our theory of neurobiological naturalism rests on three interrelated principles.

Principle 1: Life. To explain the subjective aspects of consciousness, we must factor in the many features that the brain *shares* with all living things. These are the various aspects of life that Mayr identified as distinguishing it from nonlife, and all of them are now universally accepted as consistent with standard physics and chemistry. The unique features of consciousness are in fact fully grounded in the unique features of life, so if we are to properly understand consciousness, we need to recognize it as a living process.

Principle 2: Neural features. While consciousness is built on the general features of life, it also depends on additional neurobiological features that are unique to neural systems. These are simple as well as complex reflexes and motor programs that are created at the level of the spinal cord and core brain.[9] They are not conscious but are nonetheless critical to the creation and the evolution of consciousness.

It is at the next level of neurobiology that consciousness appears. This level has gained a complex suite of special neurobiological features that are unique to conscious brains. While these special features are built on, depend on, and display all the general-life and reflexive functions, their contribution is required for consciousness. But just because these special features are new, numerous, complex, and revolutionary does not mean they lead to consciousness through any new kind of "physics of the brain" or by tapping into any mysterious "consciousness force." Rather, we can account for all of them as natural characteristics of certain complex nervous systems in the same way that life is a complex feature of specifically organized nonliving chemical constituents.

Principle 3: The subjective–objective barrier can be naturally explained. Once complex brains acquire these special features, there are, in addition, some unique but naturally explainable reasons why one's own brain processes are inaccessible from the subjective (first-person) point of view, and conversely why one's subjective experience is inaccessible from the observer's third-person point of view. The book's main theme is that we can attain a naturalized account of the subjective features of consciousness from these three principles. We build this three-principle theory of neurobiological naturalism as follows.

An Outline of the Book

In chapter 2, we begin our search for a naturalized account of consciousness, but rather than tackling the problem in one fell swoop, we break consciousness down into more manageable subparts. First, although scientists generally consider sensory consciousness as a single thing, here we more usefully divide it into three partially overlapping and interacting domains: *exteroceptive, interoceptive,* and *affective*. Exteroceptive consciousness involves creating sensory mental images, affective involves internal feelings of good and bad, and interoceptive falls somewhere in between. Not only can these three domains be viewed as different types of consciousness, but they also represent different forms of qualitative experience, or what philosophers call *qualia* (see chap. 2).[10] Over the course of the book, we show that these domains of primary consciousness or qualia have markedly different neural architectures but also share numerous special neurobiological features (see chap. 6).

Adding to the complexity of explaining consciousness and subjectivity, chapter 2 also shows that while the problem of the explanatory gap is often treated as a *single* question, this is an oversimplification of a far more complex problem. In fact, when we consider the three domains of consciousness, we find that each domain entails *multiple* explanatory gaps. We therefore suggest that a satisfactory naturalistic account must be able to explain all the types of primary consciousness, as well as all their diverse explanatory gaps.

Once we establish that there are multiple forms of sensory consciousness (images and affect) and that each entails multiple explanatory gaps, the next question is: what are the common neural features that make images and affects possible? To answer this question, chapters 3 through 5 consider *which* of the living animals have primary consciousness and *what* are its shared neurobiological bases across all these animals.

Chapters 3 and 4 focus on the vertebrates, the best-studied group of animals. Chapter 3 examines the creation of exteroceptive sensory *images* and finds both neural diversity and commonality across the vertebrate species in the creation of these mental images. Similarly, in chapter 4, when we consider how the vertebrates create conscious *affects*, we find diversity as well as shared neural patterns across all the species. We find the features of affective and image-based consciousness have many similarities, but also some differences.

To extend our analysis to all animals that might have sensory consciousness, chapter 5 investigates image-based and affective consciousness in the invertebrates by looking for the same neural features that mark consciousness in the vertebrates. Indeed, we find many of the requisite features in arthropods (e.g., insects) and cephalopods (e.g., octopuses), despite their brains being quite differently constructed. These invertebrates in turn add information to the total set of special neurobiological features needed for consciousness. By this point, we have recognized consciousness to be quite diverse, present in extremely different groups of animals, and involving different brain structures.

As noted earlier, these findings lead us to conclude that, despite the differences in the neural architectures for images versus affects *within* a single conscious brain, and despite the vast differences in the neural architectures for images and affects *across* species, there are many common factors for the creation of *all forms of consciousness and their attendant explanatory gaps*.

Therefore in chapter 6 we discuss these commonalities in depth and formalize a list of the special neurobiological features shared by all conscious animals, by both image-based and affective consciousness, and by all the different brain regions involved in consciousness. We produce a solid list on which future explorations of consciousness may be based, or which can be used as a checklist to tell if some newly discovered animal has consciousness. Additionally, we list the general features of life, of neural reflexes, and of the core brain functions on which the special features built consciousness. With these lists, we logically model three steps from nonconscious animals to conscious ones, in an uninterrupted line and a fully scientific framework. No gaps exist in this neural progression.

Chapter 7 is the main evolutionary chapter. Here we record the three steps to consciousness as they unfolded during Earth's history, as shown by the fossil record. Then we explain the adaptive value of having consciousness. Again, we reconstruct the steps logically as progressing continuously, but with a big jump occurring about a half billion years ago during the first big explosion of animal diversity. Once again there are no gaps in the progression, meaning none in the *time* progression.

Finally, in our concluding chapter, we synthesize these issues and explain how subjectivity naturally arises. We describe our theory of neurobiological naturalism as an integrated model, presenting it in a hierarchical and evolutionary context to account for the relationship between the general features of life, of nonconscious nervous systems, and finally of the special neurobiological features of conscious brains. Having unified the first and second principles mentioned earlier in the chapter (life and neural features), we add our third principle, which helps explain why subjective experience really is different from objective observation but can nonetheless be naturally explained.

2 Approaching the Gaps: Images and Affects

To provide a naturalistic account of the nature and origins of consciousness, we believe that concepts such as "something it is like to be," primary consciousness, and subjective experience require some refinements if they are to be properly understood and explained. In this chapter, we argue that, depending on what sort of experience is being processed, there is more than one variety of primary consciousness. Further, we find that there is more than one explanatory gap. Here we mostly consider humans and other mammals, as have most other workers who studied consciousness in the past. We will consider other vertebrate and invertebrate animals in chapters 3 through 5.

Forms of Subjective Awareness: Exteroceptive, Interoceptive, and Affective

Although researchers generally consider primary sensory consciousness to be a single type or single basic level of consciousness, we think it is more usefully divided into three partially overlapping and interacting domains within the overall unity of consciousness. These three domains are *exteroceptive, interoceptive,* and *affective*, and they can roughly be organized along an outer-to-inner, or extrapersonal-to-intrapersonal, continuum.[1]

Different authors have emphasized different domains in their efforts to explain primary consciousness. On one end, exteroceptive consciousness reflects how the brain processes and represents sensory information from the external environment. Many researchers consider the mental images that result from this form of sensory processing to be the prototypical kind of primary consciousness. For instance, Gerald Edelman called primary consciousness "the state of being mentally aware of things in the world—of having mental images in the present,"[2] and Antonio Damasio considered mental images a part of "core consciousness," a concept similar to Edelman's primary consciousness.[3] We agree that sensory images are important, though they are not the *only* part of primary consciousness. We also agree that any organism with a nervous system that translates its distance-sensory inputs into any sort of mental images has the exteroceptive type of primary consciousness.

At the other end of the spectrum of primary consciousness lie *affective states* of awareness, which entail positive and negative feelings. Many investigators, including Michel Cabanac, Derek Denton, Jaak Panksepp, and their colleagues, have focused on affective consciousness as the key aspect of primary consciousness.[4] The affective domain entails how the brain creates internal emotional (affective) states. Interoceptive consciousness, of sensations from within the body, lies in the middle and has features of both the other domains.[5]

While all three domains are varieties of primary consciousness, they differ in the way they are experienced by the subject. Nonetheless, whether one focuses on exteroceptive mental images, on interoceptive bodily experience, or on affects, all involve subjective experience, and all entail explanatory gaps between their subjective experiences and the brain. Although different authors have emphasized one or another of these domains to explain the nature of primary consciousness, we want to build a broader picture that includes all the domains, and to do so, we must look more carefully at their similarities and differences.

Exteroceptive Awareness and Sensory Mental Images

To begin with, all three domains of consciousness share certain building blocks. At its most basic level, the nervous system has many signaling cells, called *neurons* (fig. 2.1A). These neurons form chains and information-processing networks throughout much of the body, but especially in the *central nervous system* (the brain and an attached nerve cord or spinal cord). Neurons communicate with one another at special cell junctions called *synapses*; the communication is performed by signaling molecules called *neurotransmitters*. A simple chain of neurons is called a *reflex arc*, which is responsible for reflexes (fig. 2.1B). Neural reflexes are fast and involuntary and do not require consciousness for their operation. Examples include the jerk of the leg when the knee's patellar tendon is struck with a reflex hammer, or the constriction of the iris of the eye in bright light. Again, though we treat reflexes as the original building blocks of consciousness, they are not conscious; they occur even in unconscious people in a coma.

Exteroceptive consciousness entails the processing of sensory information from the external world: incoming sights, sounds, smells, and things that touch the skin. Exteroception, in other words, mostly involves sensing things from a distance, with the *distance senses*. If we consider exteroceptive processing from the

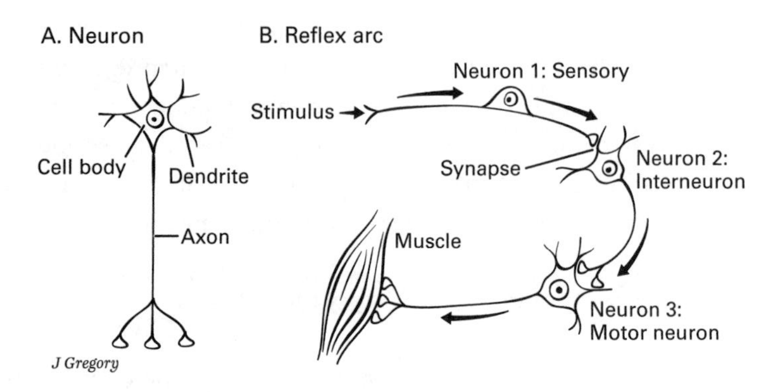

Figure 2.1
Neuron (*A*) and a simple reflex arc with three neurons (*B*). A neuron is a nerve cell. In *B*, a stimulus such as poking the skin leads to a muscle contraction that moves the body away from the stimulus. Notice the three main classes of neurons: sensory neuron, interneuron, and motor neuron. A junction between two neurons for communication is a synapse.

standpoint of subjective consciousness, the end result is that the brain builds simulations of one's surroundings called "sensory mental images." We emphasize that such images are representations of the immediate physical world and are *not* the same as *mental imagery* or *imagination*, concepts that describe the more advanced ability to form mental representations in the absence of direct perceptual input from the environment.

Exteroceptive consciousness has several specific neural features that distinguish it from affective consciousness. The major distinguishing feature of its neural pathways is that the neurons are hierarchically arranged into *topographic* or *isomorphic maps* (figure 2.2; plate 1).[6] The term *topographic map* means that an orderly representation of a sensory-receptor surface is preserved throughout the sensory pathways that go up to the brain. Some

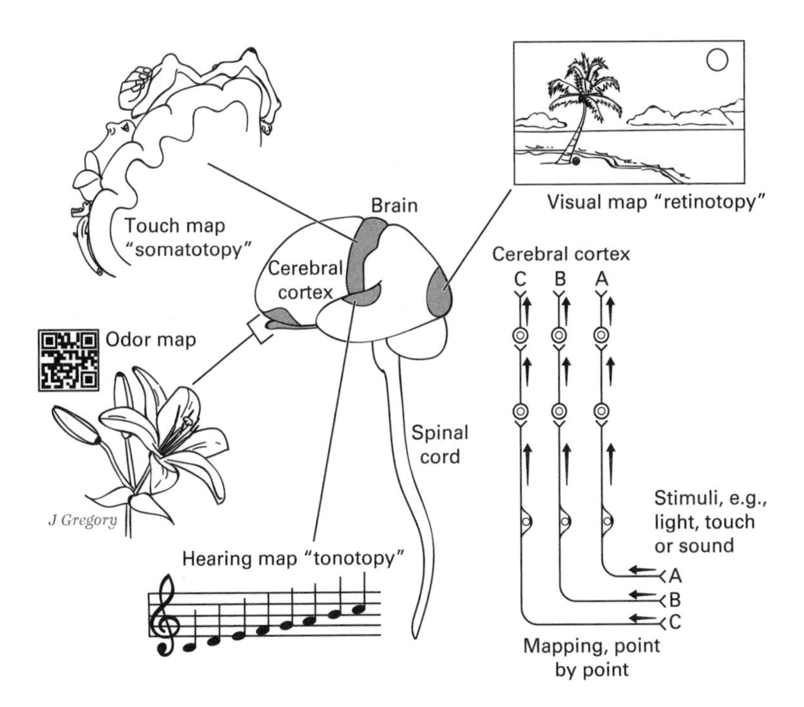

Figure 2.2

Mapped isomorphic organization of the exteroceptive sensory pathways. Each sensory pathway of several neurons (*right*) is a hierarchy that carries signals up to the brain, keeping a point-by-point mapping (A, B, or C) of a body surface, a body structure, or the outside world. This mapping leads to the mapped mental images that are drawn around the brain. The touch map of the body (*upper left*) includes a cut section through the folded cerebral cortex. The bar code associated with the flower at left shows that each complex odor has its own, coded scent signature. See plate 1.

of these maps are arranged spatially, as in the visual system, where light stimuli from the viewed world are represented point by point within the retina of the eye, and this *retinotopy* (retina mapping) is in turn preserved in the higher visual pathways in the brain. In a similar point-by-point fashion, the touch *somatotopy* (body mapping) of the surface of the skin is preserved in successive levels, so that the adjacent parts of the body are represented by adjacent neurons in the brain. In the *tonotopy* (tone mapping) of the auditory system, the vibrations of different tones of sound organize by their spatial positions on the cochlea, which is the receptor structure in the inner ear, and these mappings are in turn preserved in the higher brain levels where the sound is consciously heard. The smell path works a little differently, but it still maps things (odor combinations) into the brain and also forms a smell map of space.[7]

Affective Awareness

Another variety of primary consciousness is affective awareness. One important difference between affective and exteroceptive consciousness is that while the exteroceptive type does not *in itself* assign values to its images, affective experiences always have an intrinsic *valence*—the aversiveness (negativity) or the attractiveness (positivity) of the experience. Thus while the sight of a lion or the smell of a flower may eventually—but not necessarily—evoke fear or joy, these exteroceptive images evoke such feelings only after additional processing by different brain systems that provide the affective and inner-body aspects. Another difference is that while exteroceptive images are "local" and referred to some specific place in the world or on the body, affective consciousness is global (whole body) and involves the entire bodily individual. For example, you do not experience happiness in your foot.

The affective aspect of consciousness also relates more directly to internal motivations, drives, and behavioral responses than does exteroceptive consciousness. Positive affects such as liking and pleasure will motivate (drive) the animal to approach a rewarding stimulus, and negative affects (dislike, displeasure, discomfort) will motivate avoidance or escape from a noxious or threatening stimulus.

Finally (as addressed in chap. 4), the brain regions that produce affects are more diffuse, more numerous, and less strictly hierarchical than the brain systems that produce exteroceptive images.

Interoceptive Awareness

Interoceptive consciousness shares features with both the exteroceptive and the affective types. Interoception senses physiological and mechanical changes in the body with neuronal endings that distribute widely through the body. Like the other two forms of sensory awareness, interoception has been promoted as *the* basis of consciousness.[8] The inner-body sensations come from the viscera: the gut, heart, lungs,[9] and so on; for example, from the esophagus when you swallow hot coffee, or the feeling of air hunger in your lungs when you hold your breath. Interoception also includes some global kinds of inner sensations such as hunger, thirst, and fatigue. These diverse sensations inform the animal whether things are going smoothly inside the body so that it can adjust its internal workings to an optimal, steady state, a tuning process called maintaining the body's *homeostasis*.

Damasio and his coworkers noted a close overlap of the brain areas for interoceptive processing, for maintaining homeostasis, and for emotions (affects) in humans.[10] When compared to exteroceptive consciousness, the interoceptive type relates more directly to affects because it often has an emotional valence: experiencing a wave of nausea or feeling the pain of angina in a blood-starved heart can get one very upset very fast.

Interoception also resembles exteroception to an extent because it can form mental images of the body with some amount of somatotopic mapping. However, these interoceptive sensations are not as precisely localized as are exteroceptive ones, meaning that the somatotopic map is not as point by point. For instance, it can be hard to tell if one's abdominal pain is coming from the small intestine, appendix, gallbladder, or ovary. Nonetheless, interoceptive signals follow clear neural hierarchies of ascending sensory pathways in the brain (fig. 2.3), as do exteroceptive signals.

The neurobiologist A. D. Craig has pointed out that some kinds of exteroceptors in the skin—those responsible for pain (nociception), temperature, and itch—are necessary for homeostasis and should therefore be considered interoceptive and homeostatic.[11] All these unpleasant skin sensations have strong affective aspects, and as such they generate powerful emotions and drives. Yet they also have truly exteroceptive characteristics. For instance, a pinprick creates an exquisitely precise somatotopic image of the pin's tip as located in space, along with the unpleasant sharp sensation. We can also see pain's dual, extero–intero nature in its two subtypes, *sharp* pain versus *burning* pain.[12] Whereas sharp, pricking pain is precisely localized on the skin (somatotopically, like exteroception), the slow, dull, and

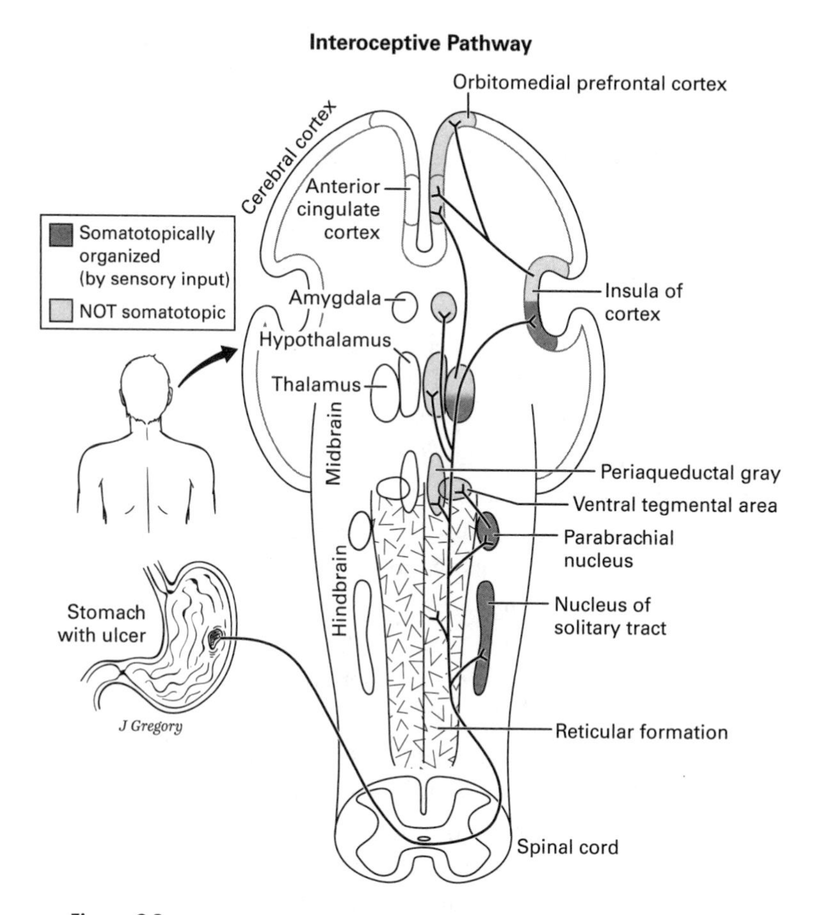

Figure 2.3

An interoceptive sensory pathway, for the pain of a stomach ulcer, running from the stomach up through the spinal cord and brain. Follow the ascending black lines. It has many branches to the brain's widespread affective regions (labeled). The pathway is hierarchical, with multiple levels of neurons, as indicated by the successive black lines. Shading shows that the amount of somatotopic (isomorphic) body mapping decreases higher in the pathway.

burning pain is less localized, covering larger areas (more crudely somatotopic, like interoception).

Adding Proprioception, a Sense of Movement

Proprioception is another class of senses that straddles the line between interoception and exteroception. Like interoception, proprioception senses inner-body structures. Specifically, it measures the amount of stretch in the body's joints, muscles, tendons, and skin as the body moves.[13] In sensing these things, proprioception informs a person or animal exactly how it is moving through space. This information lets the animal navigate smoothly, stay balanced, and target its destinations.

Many aspects of proprioceptive sensing are not conscious, instead producing automatic adjustments and reflexes. This seems to be because often we must sense and change our movements so quickly that we have no time to think about them. However, some other aspects of proprioception are consciously felt. This is especially true of joint stretch, though we can also experience some muscle stretch. To feel these things, do a full-body stretch by extending your arms and legs and notice how it feels, especially at your elbow, shoulder, knee, and hip joints. Proprioception is considered closer to exteroception than to interoception, because like exteroception it has a precise somatotopy, and because it follows some of the exteroceptive neural pathways to the brain (along with having some paths of its own). Some investigators who study consciousness focus on the proprioceptive and movement aspects because they center their theories on how consciousness directs actions and behaviors.[14] Although we focus on the sensory and experiential aspects instead, we like their motor approach and wholeheartedly agree

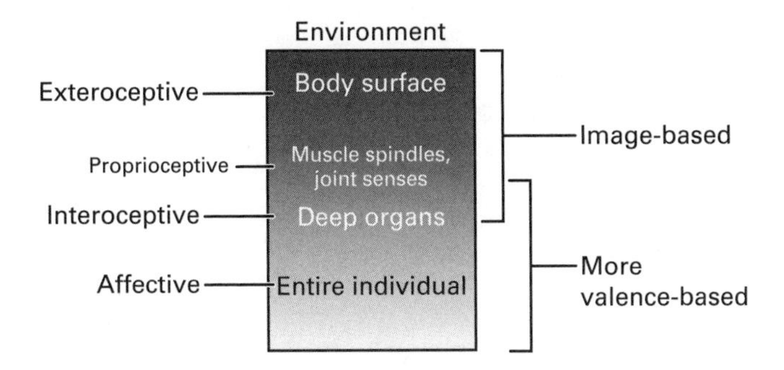

Figure 2.4
The subtypes of consciousness. *Left side*: Three domains of sensory consciousness shown as a continuum: exteroceptive, interoceptive, and affective. Proprioceptive consciousness is also shown, with less emphasis. *Right side*: In some parts of this book, we simplify the domains down to just two: image-based and affective (valence-based). Image-based consciousness includes exteroceptive, proprioceptive, and the somatotopically mapped aspects of interoceptive. Muscle spindles and joint senses measure proprioception, the stretch produced in our body's muscles and joint capsules as we move. See plate 2.

that directing movements through complex space is an important biological function of consciousness (chaps. 3–8).

Figure 2.4 summarizes the domains of consciousness. It also shows how, for brevity, we sometimes use a simpler division into image-based and valence-based aspects, or images and affects.

Multiple Explanatory Gaps

While the challenge of primary consciousness and the explanatory gap is often treated as a single question, when we consider the three domains of primary consciousness, we find that each of the three actually has multiple explanatory gaps. We have grouped these gaps into four categories, called the *neuro-ontologically*

Box 2.1

The four neuro-ontologically subjective features of consciousness

- referral
- mental unity
- mental causation
- qualia

subjective features of consciousness (*NSFCs*). They are *referral*, *mental unity*, *mental causation*, and *qualia* (box 2.1).[15]

Referral means that sensory experiences are perceived not as if in the brain, where they are constructed, but as if in the outside world or inside the body (from the stimuli received by exteroceptors and interoceptors, respectively), or as an *affective state* of positive or negative feeling that involves the entire individual. Thus referral characterizes all three domains of consciousness as external to the brain, although the most obvious example is exteroceptive projicience (projected sensation), the projection of what is seen, heard, or smelled into the outer world. In all three domains, the subjective gap of referral lies between the brain, where the experience is actually created—where we know it actually *is*—and the place to which the experience is referred. For instance, in the case of vision, I do not experience the sight of the sunset as if on my retina or in the visual areas within my brain; rather, I experience the sunset as where the stimulus came from, out beyond the horizon. Similarly, if I have appendicitis, I experience the pain in my abdomen, not in my brain. Indeed, no brain activity of any sort, whether referred to the world, to the body, or even as an emotional state, is ever experienced as occurring within the brain: experience is never referred to the neuronal circuits that cause it. Explaining this feature of subjectivity is central to the theory of the nature of consciousness that we present in the final chapter.

In *mental unity* the gap appears between the divisible, discontinuous brain that consists of individual neurons and the unified, continuous field of awareness. This unified stage of consciousness integrates many things into one experience. For example, it integrates all these: a sensory simulation of the surrounding world, of one's location in that world, of one's present mood, and of inner-body conditions such as the awareness that one now has a cold and respiratory infection. This unity gap is called the *grain problem*, meaning that the divisible "grain" of the neurons of the brain is far coarser than the apparently seamless grain of subjective experience.[16]

Mental causation is the puzzle of how the subjective, seemingly immaterial, and objectively unobservable mind can cause physical effects in the material world. These physical effects include consciously signaling the muscles and movements of one's own body, as well as the effects these actions cause on material objects in the world.[17] For example, in the days after people in a town experience a traffic-paralyzing snowstorm, hunger and worry about being homebound attract them to all the food in their refrigerators, *causing* the fridges to empty.

Finally, *qualia* are perceived qualities of objects in the outside world or inner body or of one's affective states, such as emotions. Viewed colors, the feeling of a full stomach, and the experience of anger are all qualia. As we shall see, explaining qualia is considered by many philosophers to be the holy grail of explaining consciousness, the particular explanatory gap that is given the most attention and appears the most mysterious of all the subjective features of consciousness.[18] Philosophers typically pose the following questions about qualia: Are there specific neurobiological features that create qualia? Why do certain neural states feel the particular way they do? Why does the brain even produce feelings at all?

In summary, in this chapter we have established the three partly overlapping domains of sensory consciousness (figure 2.4), each of which involves four explanatory gaps, a gap for each subjective feature of consciousness (box 2.1). The next step in naturalizing consciousness is to explain how the gaps arise. We approach this problem by asking which animals are conscious, that is, are capable of having images and affects.

3 Naturalizing Vertebrate Consciousness: Mental Images

As we discussed in chapter 1, according to Nagel, if an animal has "something it is like to be," then it has primary consciousness. Some might argue that this is a fairly low bar to set for something as complicated and important to human affairs as consciousness. But since we are trying to determine whether an animal has primary consciousness—the most basic form of any experience—then this is a reasonable criterion.

We have argued that the brain's construction of mental images and affects is pivotal to the creation of primary consciousness and the explanatory gaps. But to naturalize consciousness, we need to show that these subjective experiences do not somehow magically emerge from the brain out of the blue. To do so, in chapters 3 through 5, we consider which of the living animals have primary consciousness, and we deduce its neurobiological basis, so as to address later (in chap. 6) what common biological and neurobiological features make consciousness possible.

In this chapter and the next, we focus on the vertebrates—fish, amphibians, reptiles, birds, and mammals—because this is the group to which humans, the only animals known to be conscious with certainty, belong. In this chapter, we ask which vertebrates have image-based consciousness, mostly of the exteroceptive type, and we compare the relevant neurobiology across species. In chapter 4 we do the same for affective consciousness. In chapter 5 we turn our attention to invertebrates.

Which Vertebrates?

Most people who study consciousness have gone beyond the long-held notion that only humans have primary consciousness. Many now assign consciousness to all mammals and birds,[1] both of which have an expanded dorsal pallium of the forebrain. In mammals, this is the cerebral cortex, and the central idea is that the cerebral cortex (or the equivalent bird pallium) is responsible for consciousness, as it works together with another part of the forebrain called the thalamus.[2] One line of evidence for this idea is that damage to the corticothalamic system in humans leads to a loss of conscious sensations and images.[3] This finding offers convincing evidence for *humans* and other mammals, but we should also consider the other groups of vertebrate animals, whose cerebrum lacks a cerebral cortex (fig. 3.1). Even fish appear conscious, because of their elaborate senses; their advanced, hierarchical sensory anatomies (including sharp vision); and the alert attention they pay to stimuli.[4]

Figure 3.1
Brains of various vertebrate animals, which we deduce can create exteroceptive mental images. Notice the relative sizes of the cerebrum and the optic tectum in the different animals. Only mammals have a cerebral cortex. We included two different mammal brains (*E*, *F*) so that one could be cut in half lengthwise to show its inner structures (*F*). The brain stem, which contains the reticular formation for conscious arousal, consists of the medulla, pons, and midbrain. See also fig. 4.1.

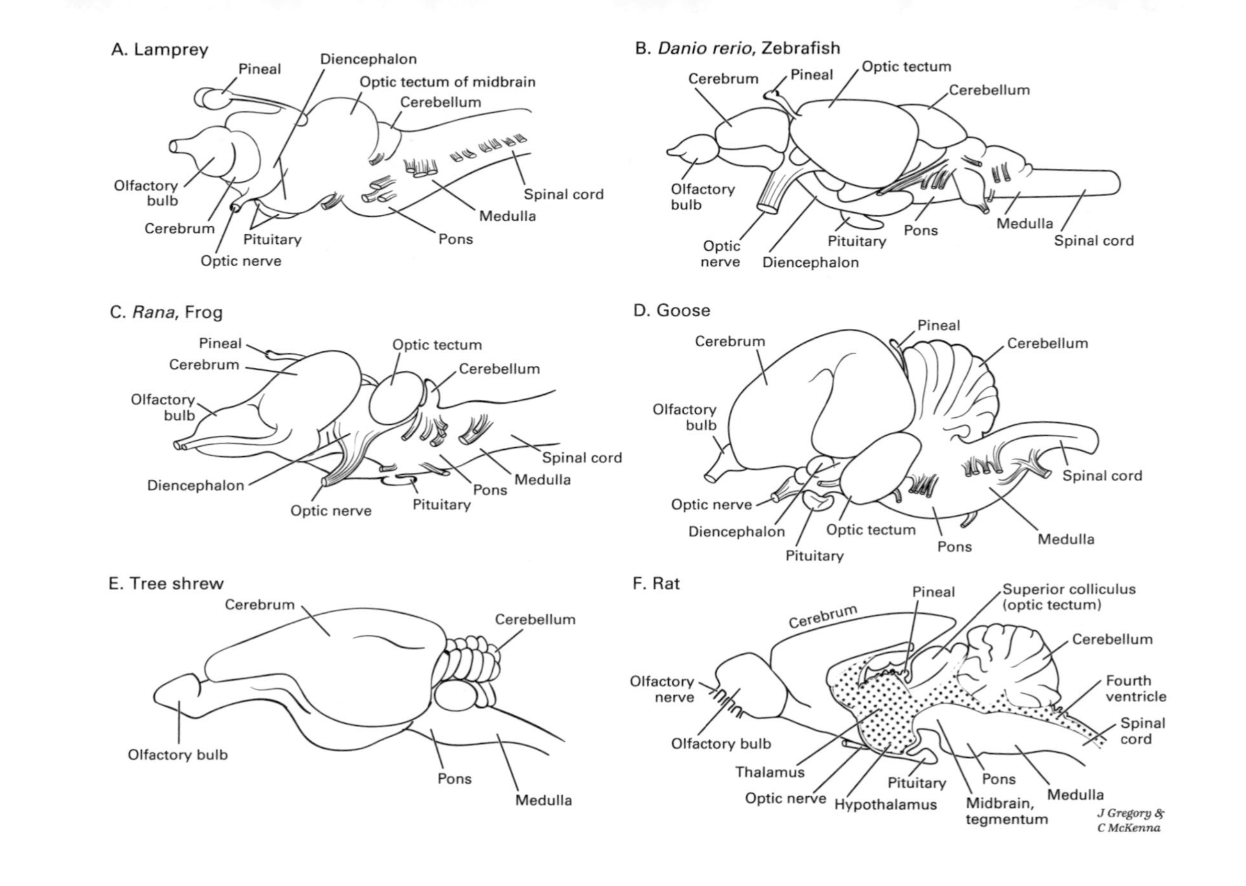
A. Lamprey
Pineal
Diencephalon
Optic tectum of midbrain
Cerebellum
Olfactory bulb
Cerebrum
Pituitary
Optic nerve
Pons
Medulla
Spinal cord
B. *Danio rerio*, Zebrafish
Cerebrum
Pineal
Optic tectum
Cerebellum
Olfactory bulb
Optic nerve
Diencephalon
Pituitary
Pons
Medulla
Spinal cord
C. *Rana*, Frog
Pineal
Cerebrum
Olfactory bulb
Optic tectum
Cerebellum
Diencephalon
Optic nerve
Pituitary
Pons
Medulla
Spinal cord
D. Goose
Cerebrum
Pineal
Cerebellum
Olfactory bulb
Optic nerve
Diencephalon
Pituitary
Optic tectum
Pons
Medulla
Spinal cord
E. Tree shrew
Cerebrum
Cerebellum
Olfactory bulb
Pons
Medulla
F. Rat
Pineal
Superior colliculus (optic tectum)
Cerebrum
Cerebellum
Olfactory nerve
Olfactory bulb
Fourth ventricle
Spinal cord
Thalamus
Optic nerve
Hypothalamus
Pituitary
Midbrain, tegmentum
Pons
Medulla
J Gregory & C McKenna

Therefore, since we want to elucidate the most basic neurobiological underpinnings of sensory consciousness, it makes sense to look at the more basal species.

Based on our reasoning that any brain that can create mapped sensory images has exteroceptive primary consciousness, we looked for the most basal vertebrates that have the neural architecture for isomorphically mapped, multisensory representations.[5] Somewhat surprisingly, this turns out to be a feature of *all vertebrate brains*. In mammals and birds, we find the main isomorphic maps in the cerebral cortex (mammals) or the correspondingly enlarged parts of the cerebrum (birds), although they appear in different cortical locations in these two animal groups (fig. 3.2).[6] However, in every vertebrate, a midbrain structure called the *optic tectum* contains a finely detailed, point-by-point map of the sensed external environment, mostly from visual inputs, but also from the hearing, touch, and balance pathways (fig. 3.3; plate 3).[7] This implies that all vertebrates—even the earliest to have evolved—have the ability to create mapped *mental images*, and thus we deduced that all vertebrates have exteroceptive primary consciousness.[8]

The optic tectum makes up a relatively large part of the brains of fish and amphibians (figs. 3.1, 3.3). We deduced it to be the site of image consciousness in these animals. In all the vertebrates, its map lets the animal attend to, look at, and turn toward the most important object in the visual field.[9] In fish and amphibians, the tectum is said to be for object "recognition" and "perception."[10] These are terms of consciousness, so our claim for tectal consciousness is not merely based on anatomical mapping. It follows, therefore, that the brain site of image-based consciousness shifted from the tectum of more basal vertebrates to the cerebral cortex during the evolution of mammals.[11]

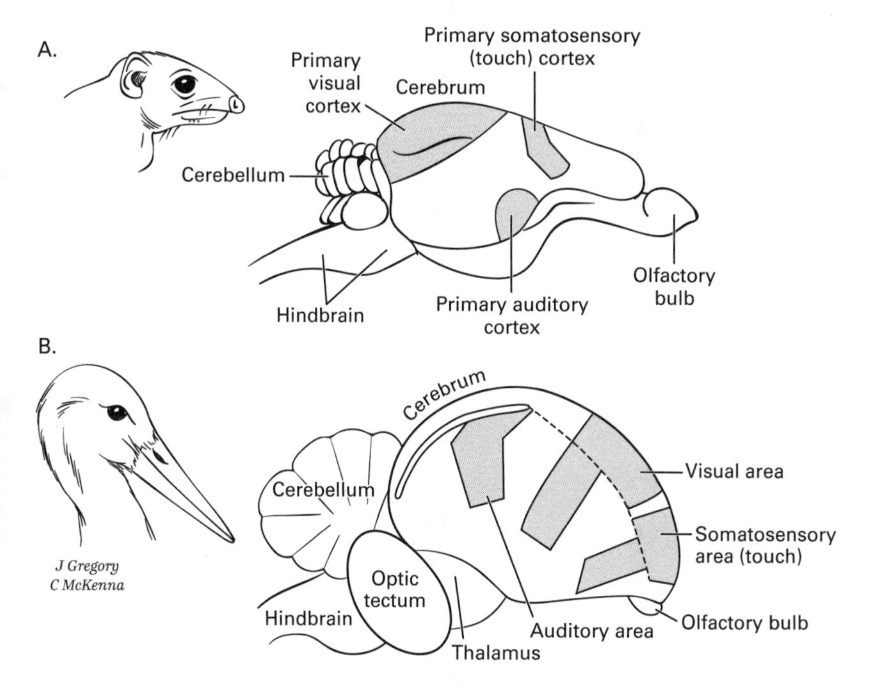

Figure 3.2
Primary sensory areas for conscious mental images in the cerebrum of a mammal (tree shrew) (*A*) and a bird (stork) (*B*). All these areas are isomorphically mapped, but they have different cerebral locations. In mammals the arrangement, from back to front, is visual, auditory, then somatosensory, whereas in birds it is auditory, visual, then somatosensory. The different arrangement suggests that these mapped sensory areas evolved independently in mammals versus birds.

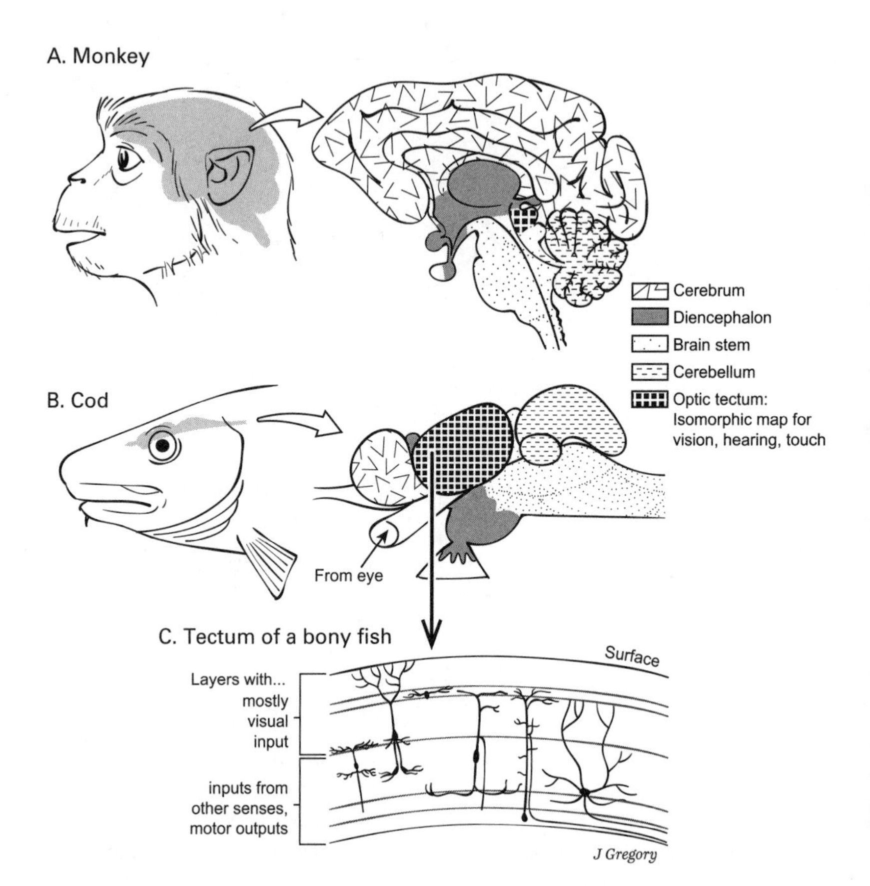

Figure 3.3
Optic tectum of the brain of a mammal (*A*) and a bony fish (*B*), with a section through the fish tectum (*C*). *C* shows some of the kinds of neurons in the tectum. We make the case that the fish tectum forms conscious images, but the mammal tectum does not. Brain in *A* is cut in half lengthwise; brain in *B* is uncut and whole. See plate 3.

Brain scientists are just starting to study the tectum at the level of its sense-integrating neural circuits, but early findings hint that it is complex enough to form highly processed images for directing behaviors in space. The tectum contains three or more layers of densely packed neurons (fig. 3.3C), of up to fifteen different neuron types (goldfish), and along with its many sensory inputs, it has extensive connections to both lower and higher centers of the brain that control actions.[12]

However, we are not claiming that the tectum performs *all* the tasks of image-based consciousness in fish and amphibians. The tectum does not perform *smell* perception, as it receives no direct input from the smell pathway. The pallium of the brain's cerebrum performs that function instead. This means smell consciousness has always involved the pallium, from the first vertebrates to mammals. Also, the pallium—not the tectum—is the location of the main memory-forming part of the vertebrate brain, named the *hippocampus* (see fig. 4.1). Using the hippocampus to recall memories is an essential part of consciousness: remembering and reexperiencing the look of a previously encountered predator, for example, permit instant recognition and escape.[13]

Furthermore, the tectum does not determine the *level* of consciousness, the amount of *arousal* that establishes how alert the animal is to stimuli.[14] All vertebrates use the same, nontectal, brain regions for arousal, especially the reticular formation in the brain stem and parts of the basal forebrain (fig. 4.1). Neurons of these regions project widely throughout the brain and release chemicals for arousal, alertness, and vigilance.[15] To summarize, smell, memory, and arousal are parts of consciousness in which the "conscious" tectum is not primarily involved.

The optic tectum does play a role in *selective attention,* which is vital for consciousness.[16] Selective attention is important because we are most conscious of those stimuli to which we attend. As mentioned earlier, the tectum finds out which object in viewed space is the most important (salient), and it directs the eyes to look toward that selected object. For this process, scientists formerly believed that the tectum had to team up with nearby centers in the brain stem (called the isthmus nuclei). However, a recent study of a basal fish, the lamprey, indicates that the tectum can do it alone—and in an extremely interesting way (fig. 3.4; plate 4).[17] When an object in the lamprey's environment emits two different kinds of sensory signals (e.g., visual signals and electric fields), then both these inputs *converge* on the same neuron in the tectal map, and that neuron signals the eyes to look at the specific point in space from which the two signals arose. The study thus not only showed that the tectum can direct attention by itself but also emphasized the tectum's role as a brain region where multiple senses come together.

Micheal L. Woodruff says that the type of attention most likely to reveal a fish has image-based consciousness is persistent visual search for an object in space, such as a food item or a mate. He has described a study by Mor Ben-Tov and her colleagues on archerfish, which seek and find insect prey in the foliage above their estuary water, then shoot down this prey with jets of water. The link to consciousness is that the archerfish must have a mapped mental image of both the background scene and of the attended prey item against that scene. And Ben-Tov found that the neurons responsible for this distinction are in the archerfish's optic tectum. This is how Woodruff traces conscious attention to the tectum in a jawed fish.[18]

Our past work focused heavily on the value of lampreys for revealing the origins of consciousness (fig. 3.4).[19] Lampreys are jawless fish in the oldest surviving group of vertebrates. A lamprey has the smallest brain, per body size, of any vertebrate,

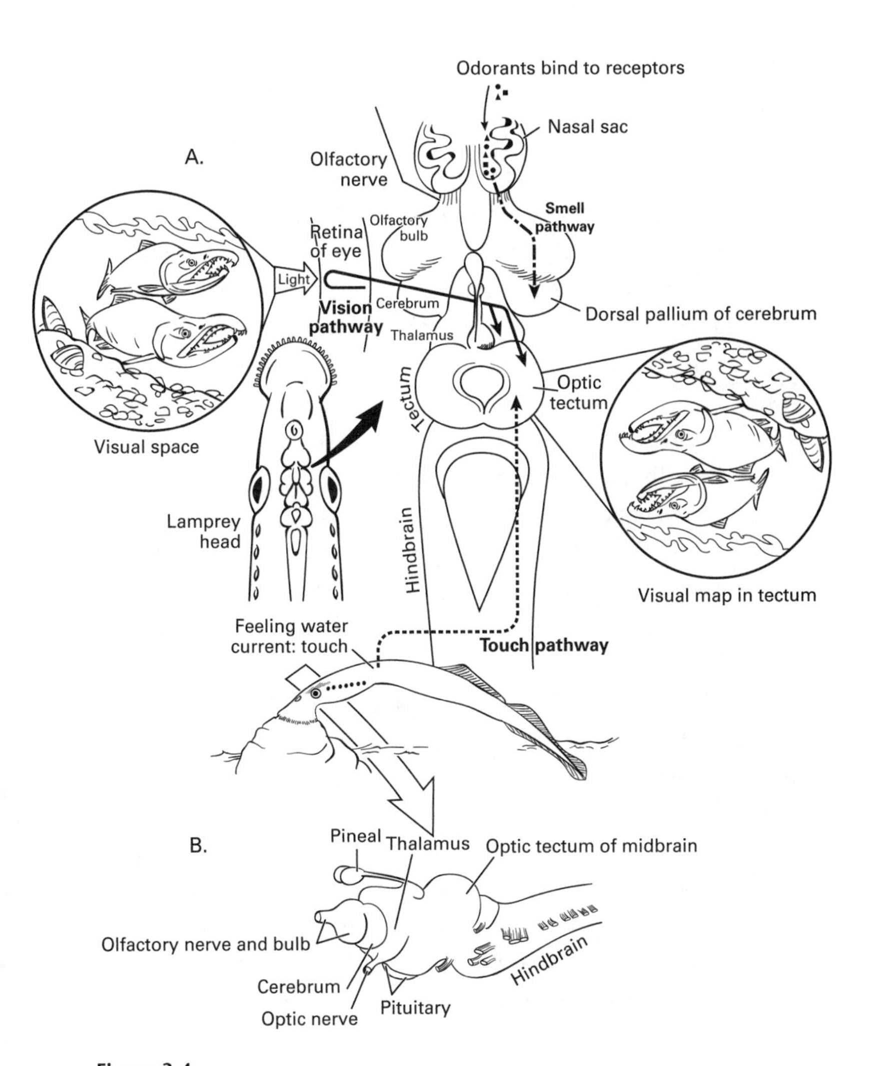

Figure 3.4

Sensory pathways and the brain of the lamprey, a jawless fish. A lamprey is shown sucking on a rock in a streambed, near the middle of the figure. Brains are viewed from above (*A*) and from the side (*B*). As shown in *A*, most sensory pathways reach the optic tectum, where they form a mapped (isomorphic) representation. This image shows two salmon in a spawning run in the same stream. See plate 4.

yet it has all the complex sensory organs of vertebrates, from excellent eyes for vision to sensitive touch, keen smell, hearing, and the typical fish sense for detecting electric fields in the water. The lamprey brain has a perfectly respectable optic tectum, with the same type of mapped hierarchical input of multisensory information as in any fish brain. Thus if fish have image-based consciousness, then lampreys do too.

The next big question is, are these tectal sensory representations in lampreys and other fish really conscious? To this we say yes. Just like all sensory images, including those experienced by humans, lampreys' images meet the criteria of the neuro-ontologically subjective features of consciousness (NSFCs) we presented in chapter 2, and hence of subjectivity. First the images are *referred* to the environment. Put another way, the lamprey responds appropriately to complex visual stimuli "in the environment." Second, the images are *unified*. Lampreys have been called "vampire killer eels," and the way they choose, approach, and attach to the swimming fish that they parasitize also suggests that the size, shape, movement, color, and so on, of these environmental targets are "bound" into unified sensory images in the lamprey brain.[20] Third, these images are *causal*. By this we mean that a lamprey can respond to environmental stimuli in complex, prolonged, planned, purposeful, nonreflexive ways that cause material changes in the external world via movements: for example, lamprey mating behavior is complex and involves moving rocks to build a nest in the streambed.[21] Finally, the sensory images are *qualitative*: by definition, consciousness recognizes different qualia as distinct from one another, and fish can perform high-level sensory discriminations based on sounds, visual characteristics, electrical signals, smell, and taste.[22]

Hence we argue that the sensory images of fish possess all four of the NSFCs we associate with subjectivity, and by these criteria they are creating conscious mental images. In other words, there is "something it is like to be a lamprey." It then follows, to echo Nagel's famous question, that surely there is something it is like to be a bat.

To summarize this section, we began with the basic deduction that mapped, multisensory representations in the brain indicate mental images, then used that deduction to reason that all vertebrates have image-based consciousness. Given this foundation, we can start to look for additional shared features of consciousness, by comparing and finding the commonalities between the conscious sensory pathways of mammals and lampreys (figs. 2.2 and 3.4). The shared features we have identified so far include having many complex sensory receptors (in eyes, nose, etc.), a brain, sensory pathways arranged in hierarchies that are at least three neurons long,[23] a place in the brain where the multiple senses come together, and brain locations for arousal and directing attention. We identify more such features in upcoming chapters.

The Invertebrate Cousins of Vertebrates

From the features we have just deduced, it follows that animals that lack primary consciousness must have less-elaborate sense organs and brains and use reflexive behaviors and simple movement programs that do not need a unified image of the world. Let us look at the closest living relatives of vertebrates to see whether they qualify as nonconscious or conscious. These relatives, all of which live in the ocean, are the fish-like amphioxus (or cephalochordates) and the bag-like tunicates (or sea squirts) (fig. 3.5). The fossil record shows that they and the vertebrates split into separate lineages before 520 million years ago, during the Period of Earth history called the Cambrian. The

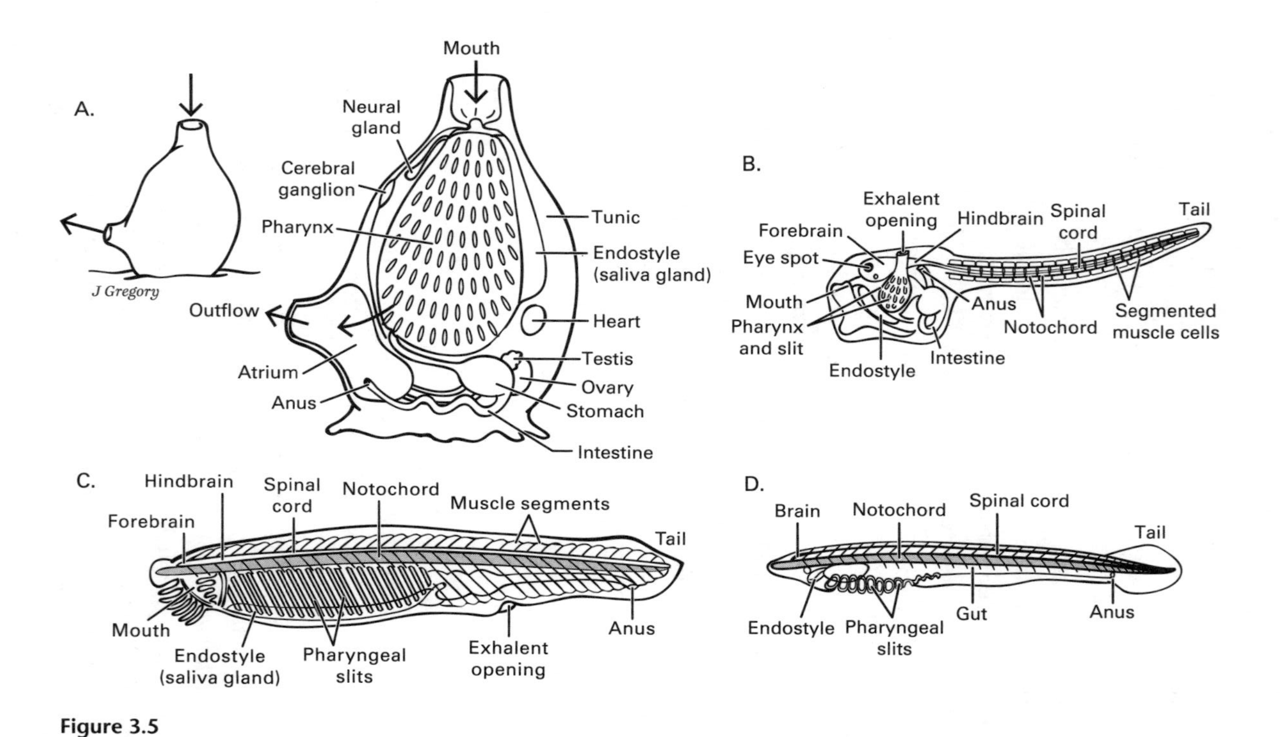

Figure 3.5
The closest invertebrate relatives of the vertebrates. Tunicate adult (*A*) and young larva (*B*); amphioxus adult (*C*) and larva (*D*).

neurobiologist Thurston Lacalli has studied the amphioxus and tunicate nervous systems for decades, and his work provided much of the following information.[24]

By our criteria, the nervous systems of tunicates and amphioxus lack the complexity required for consciousness. Their sensory equipment is not up to the task: they have simple light receptors, mechanical touch receptors, and possibly chemoreceptors, but not the camera eyes, ears for sound detection, or odor-smelling noses of vertebrates. Tunicate brains, or cerebral ganglia, are small, with several hundred neurons at most. In amphioxus the sensory pathways to the small brain, although only studied in young larvae, have just one or two levels of sensory neurons and seem to belong to reflex arcs rather than to long hierarchies of information processing (fig. 3.6). As shown in figure 3.6, light detection and mechanical touch are the main senses, and they arouse the animal to swim by signaling a motor center.[25] Researchers have not found a pallium or tectum in the amphioxus brain.

Vertebrates from Invertebrates

The finding that tunicates and amphioxus lack the necessary features implies that image-based consciousness evolved at the dawn of the vertebrates. What explains why vertebrates have so many features of consciousness, while tunicates and amphioxus have none? We have related this phenomenon to the appearance of the camera eye in the first fish,[26] namely to the vertebrate eye, because such an eye forms a photograph-like image on its retina, detailed and complete, which, when processed by an elaborating optic tectum, becomes a mental image of visual space. After this visual image evolved, information from the other senses (for

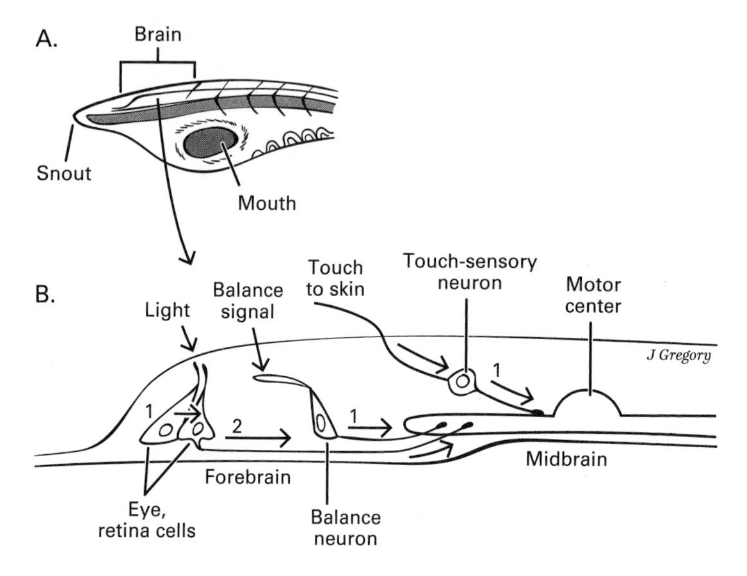

Figure 3.6
Larval amphioxus, showing the sensory pathways to its simple brain. *A* is the animal's head region; *B* is its brain. The sensory pathways are just one or two neurons long before they reach the motor center to influence motor actions. So few neurons allow little processing, so this animal probably is not conscious.

touch, sound vibrations in the water, and detecting electrical fields) was integrated into the newly evolved visual map.

But the camera eye that triggered these events requires a focusing *lens* to work well, and the lens of the vertebrate eye develops from embryonic structures that are unique to vertebrates, called *ectodermal placodes* in the outer surface layer of the body. In fact, these placodes, plus another vertebrate-only embryonic tissue called the *neural crest*, develop into all the special, conscious-associated sensory structures that distinguish vertebrates from amphioxus and tunicates. These are the sound- and smell-sensing cells, and more (fig. 3.7).[27]

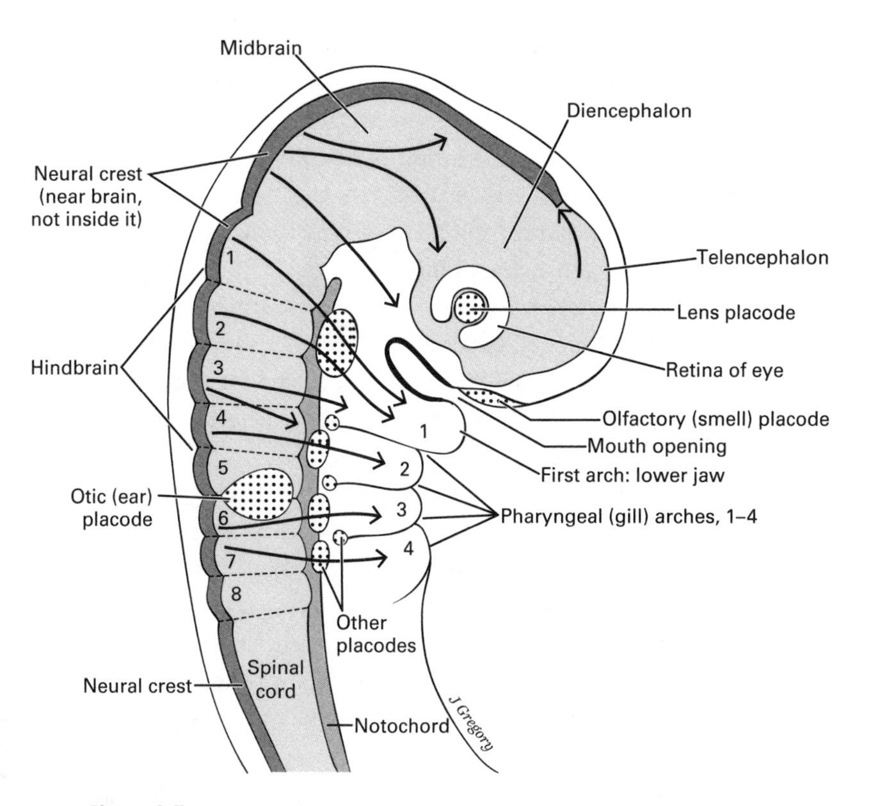

Figure 3.7

Neural crest and placodes in the head region of a vertebrate embryo. These structures are essential for sensory consciousness. Placodes are stippled; neural crest is dark gray. The long arrows show that the crest cells migrate widely in the embryo's body.

This evidence suggests that a revolution occurred near the dawn of the vertebrates, a revolution in which all the senses were refurbished and upgraded. The sensory jump brought dramatic advances in the brain as the inflow of new information arriving from the placode and crest senses of smell, hearing, improved touch, and so on, joined the flood of visual information from the newly elaborated, image-forming eyes. To process all the sensory information for image-based consciousness, the optic tectum and cerebral pallium appeared, as enlargements of pre-existing parts of the brain. The pallium evolved for consciously perceiving smell, and the tectum evolved for perceiving vision and all the other senses.

To summarize the chapter, isomorphic maps are the cornerstone of image-based sensory consciousness, these maps evolved in early vertebrates more than 520 million years ago, and this process was the natural result of the extraordinary innovations of the camera eye, neural crest, and placodes. These events led to the mental images that mark the creation of the mysterious explanatory gaps and the subjective features of consciousness outlined in chapter 2.

But while isomorphic maps are fundamental to the creation of all types of exteroceptive sensory consciousness, we find considerable diversity across species regarding exactly which parts of the brain (e.g., tectum versus cortex) can create them. In the next chapter, we explore whether this commonality within diversity is also true for *affective* consciousness.

4 Naturalizing Vertebrate Consciousness: Affects

Which Vertebrates Feel Affects?

While some investigators believe that mapped mental images hold the key to consciousness, others focus on the creation of affects, the most basic of which are feelings with valence: positive or negative conscious feelings.[1] If we are to compare the neurobiology of affective with exteroceptive consciousness, and to decide what fundamental neurobiological features of affects exist within and across species, then the first question we must confront is: which vertebrates feel affects?

This type of consciousness does not have the neat topographical maps by which we have identified image-based consciousness, so we must find a completely different marker that shows that an animal experiences affects. Indeed, we have found a whole set of these markers in some behaviors that can be considered to reveal positive and negative affects, feelings of liking and disliking (box 4.1).[2] These behaviors had to involve more than just approaching helpful things or avoiding harmful things, because even the nonconscious one-celled organisms like bacteria do that. No, each behavioral criterion goes beyond these

Box 4.1

Behavioral criteria showing an animal has affective consciousness (likes and dislikes)

1. Global operant conditioning (involving whole body and learning brand-new behaviors)
2. Behavioral trade-offs, value-based cost-benefit decisions
3. Frustration behavior
4. Self-delivery of pain relievers or rewards
5. Approach to reinforcing drugs or conditioned place preference

simple responses in requiring remembered affects, extended affects, or several different categories of affects.

The first criterion was that the animal shows a kind of associative learning called *operant conditioning*. Operant conditioning is learning to associate a behavior with a consequence of that behavior, such as learning to press a bar to obtain a food pellet.[3] By contrast, *classical conditioning*, or *Pavlovian conditioning*, is learning in which a conditioned stimulus (CS) such as the sound of a bell becomes associated with an unconditioned stimulus (UCS) such as the presence of food, until the animal responds to the CS alone (by salivating, for example). Operant conditioning is considered the more complex and advanced of the two. It entails learning from experience, and it requires more trial and error, more brainwork and memory, than does classical conditioning, where the learning is easy because the UCS is such a reliable reminder for the CS (in classical conditioning, the bell *always* means the food is coming). Accordingly, operant conditioning is more likely to stem from learned valences and memories of true affective states.

We actually choose *global* operant conditioning, which involves learning a brand-new behavior that applies to the whole body. This goes beyond just learning to change the movements of a single body part and beyond merely learning to switch from one rigidly preprogrammed behavior to another.

Recently, a more sophisticated form of learning called *unlimited associative learning* (UAL) was proposed as a marker of whether an animal has affective consciousness (and consciousness in general). Bronfman, Ginsburg, and Jablonka argue that whereas simple operant conditioning can involve learning just a single characteristic of a new object, UAL involves more than this.[4] That is, UAL means learning the reward value of the *whole* object: the appearance, weight, and smell of the food pellet all together, so that the animal will never be fooled by a pebble that looks like the food. UAL also means the ability to learn all the steps in a complex response: a rat finds the food-delivering lever, reaches it, learns how hard to press it and when to stop, and so on. Indeed, this represents sophisticated affective learning of brand-new behaviors.

However, Bronfman and coworkers propose that UAL is *the* marker of the evolutionary transition to minimal consciousness. While we have no problem using UAL as *one* marker that the evolutionary transition to minimal consciousness has occurred, and we agree that UAL was an important driver of evolution, we see it as being a bit too advanced (too "unlimited") to be *the* marker. It seems that affects could have been present in animals with more-limited forms of global operant learning. That is, UAL is a feature of brains that *had already evolved* minimal consciousness.

In practice, however, both we and the UAL team came to the same conclusion. We all searched the literature of animal behavior and found that while every group of animals with a nervous system can learn by classical conditioning,[5] learning by global operant conditioning and UAL occurs only in the vertebrates, arthropods, and cephalopod mollusks such as octopuses and squid (chap. 5). By our criteria, these are the animals that possess affective consciousness. Certainly, all the vertebrate groups—fish, amphibians, reptile, birds, and mammals—meet the criteria.[6]

The vertebrates also meet our other behavioral criteria for affective consciousness (box 4.1). *Behavioral trade-offs,* such as weighing the benefit of getting more food against a risk of higher predation near the food source, indicate that two different valences are recognized and weighed against each other. Trade-offs have been documented in all the vertebrate groups. *Frustration behavior,* such as behaving aggressively after a reward is denied, shows that negative feelings or affects can last (so therefore such affects must exist). This frustration has been demonstrated in fish, birds, and mammals. The final two criteria are: *self-delivery of pain relievers or rewards*; and *approaching drugs* (e.g., amphetamines, ethanol) or *conditioned place preference,* which is preferring to be where one previously received drugs or rewards. These criteria are informative because they go beyond basic "approaching" and reach the higher level of actively seeking and pursuing the positively valenced stimuli. Again, all vertebrate groups have been shown to do these things.[7] In summary, behavioral studies indicate all vertebrates have affective consciousness.

Does the brain's *anatomy* support this conclusion? That is, do all vertebrates have the brain structures for the basic, raw affects? First, we must pinpoint these structures in humans. Some experts say that our cerebral cortex is the seat of all conscious emotions, but others say the cortex only adjusts (inhibits) the raw emotions that are generated by lower, subcortical regions of the brain (the shaded regions in fig. 4.1).[8] Good evidence for the idea of subcortical affects is that humans and other mammals whose cortex is damaged or absent show strongly emotional responses and behaviors.[9]

This phenomenon is illustrated by human children born with a condition called hydranencephaly. Their developing cerebral cortex is nearly fully destroyed by something that cuts off its blood supply before birth, though their subcortical brain remains intact. These children "express pleasure by smiling and laughter, and aversion by 'fussing,' arching of the back and crying (in many gradations), their faces being animated by these emotional states. A familiar adult can employ this responsiveness to build up play sequences predictably progressing from smiling, through giggling, to laughter and great excitement on the part of the child."[10]

This behavior suggests that at least some children experience affects without having a cerebral cortex. Indeed, in a recent survey by Aleman and Merker, a solid majority of the primary caregivers of hydranencephalic children reported emotional behaviors that go against the common belief that hydranencephaly is a vegetative state.[11]

Additional evidence that affects can arise subcortically comes from electrically stimulating the subcortical regions with electrodes in a surgical procedure called *deep brain stimulation* (DBS).[12] DBS stimulates the medial forebrain bundle (MFB), a bundle of nerve fibers that interconnects many of the affective regions (fig. 4.2). Upon stimulation of one part of the MFB, humans feel negative emotion (fear, rage, panic), and rats routinely try to escape, as if experiencing the same negative emotion. And

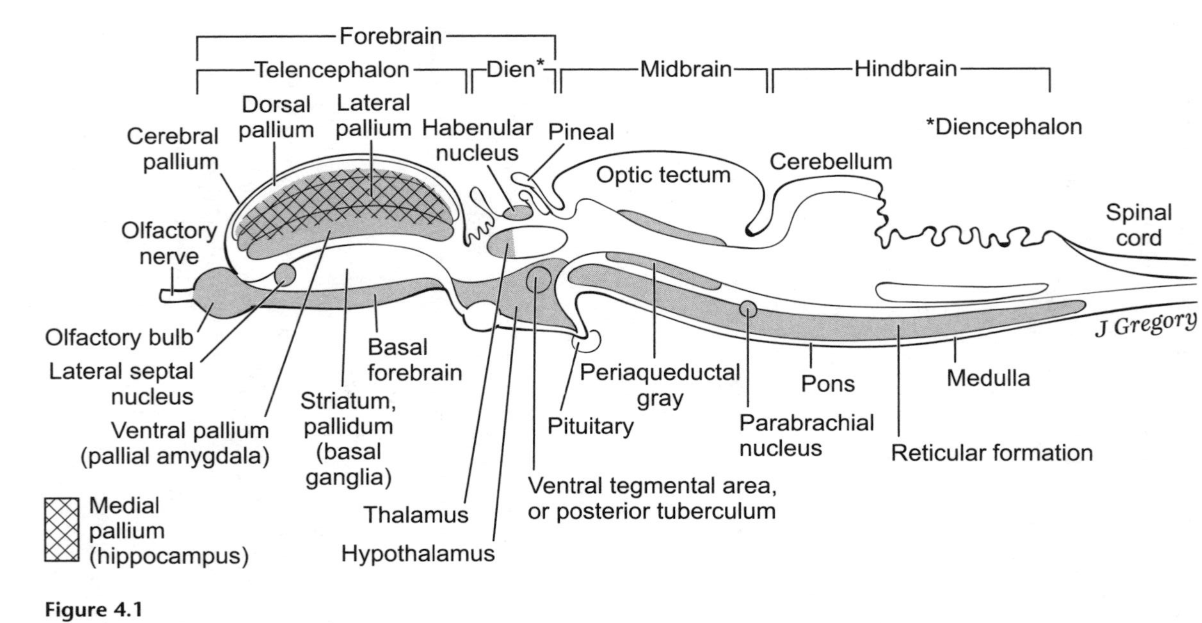

Figure 4.1
Regions of the vertebrate brain for raw affects or basic emotions are shaded. Dorsal pallium, which is the cerebral cortex of mammals, is not shaded. Many of these structures are also shown in figure 2.3.

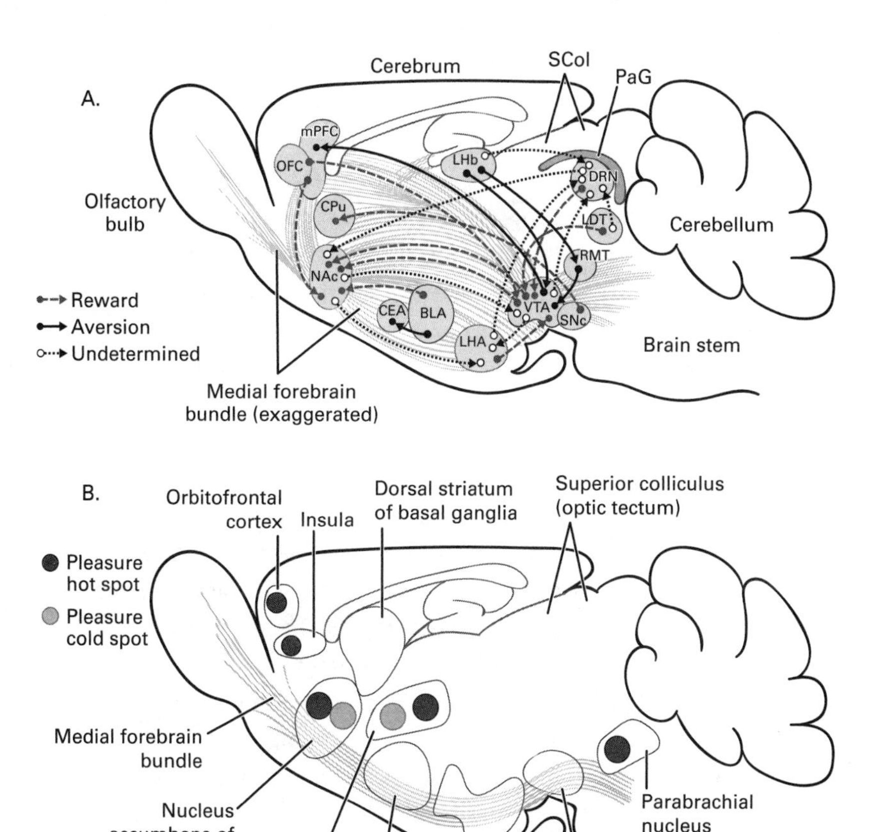

Figure 4.2
Regions of the rodent brain that associate with affects (positive and negative emotions). For orientation, see figure 3.1F. Both parts of the figure show the medial forebrain bundle of fibers (*light gray*), which interconnects the affective regions. (*A*) shows the major affective centers and their interconnections by valence neurons that code reward or aversion, from mouse studies. (*B*) shows the pleasure hot spots (*black dots*) and cold spots (*gray dots*) in some of the affective regions of the rat brain. These spots were found by applying opioid drugs to these regions and then seeing if this increased or decreased the rat's facial expressions of pleasure when it tasted sugar. Part *A* is modified from fig. 3 in Hu 2016, and part *B* is from fig. 1 in Berridge and Kringelbach 2015. Key for part *A*: BLA: amygdala (basolateral part); CEA: amygdala (central part); CPu: caudate and putamen parts of the basal ganglia; DRN: dorsal raphe nucleus of reticular formation; LDT: laterodorsal tegmental nucleus of reticular formation; LHA: lateral hypothalamus; LHb: lateral habenular nucleus; mPFC: medial prefrontal part of cerebral cortex; NAc: nucleus accumbens of the basal ganglia; OFC: orbitofrontal part of cerebral cortex; PaG: periaqueductal gray; RMT: rostromedial tegmental nucleus of reticular formation; SCol: superior colliculus (optic tectum); SNc: substantia nigra pars compacta of the basal ganglia; VTA: ventral tegmental area.

upon stimulation of another part of the MFB, humans feel positive emotion (caring, playfulness, lust, a desire for curious exploration), and rats self-activate the DBS device, indicating they likewise experience positive emotion and want to continue the experience.[13] All this evidence indicates that every mammal has affective consciousness, for which specific subcortical brain structures are responsible.

This raises the question of whether other vertebrates besides mammals have the same brain centers for affects. To find an answer, we must take an anatomy-only approach, because no functional brain-stimulation studies have been performed in the nonmammal vertebrates. We searched the literature for the affective brain structures and found them to be highly conserved across the vertebrates. Even fish have most of the affective structures: lampreys have at least 19 of 25, and bony fish have 23 of 25. That is a summary of our earlier findings,[14] and here we can add that all groups of vertebrates including lampreys have a medial forebrain bundle, the key structure that interconnects so many of the affective regions of the mammal brain (fig. 4.2).[15] We conclude that all vertebrates have the brain structures for affective consciousness.

Affective Circuits

In chapter 3, when we discussed image-based consciousness, we could readily see how the neural circuitry might lead to the experience: it was by building topographically mapped simulations of the world and body. Can we also link circuitry to experience for affective consciousness? Here it is not so easy, because the affective circuitry largely lacks topographic mapping. Also, the brain's affective centers are more numerous and spread out than those of the exteroceptive pathways for images, and they receive more branches from the ascending axons (fig. 2.3). The intercommunicating axons of affective pathways also branch a lot more than in the exteroceptive pathways, sending signals to many different parts of the system. Another difference is that affective circuits communicate less through short-distance neurotransmitter chemicals and more through far-diffusing *neuromodulator* chemicals than do exteroceptive circuits.[16]

While these distinctions may seem fuzzy, recent studies in rodents provide more concrete information about affective structures. Individual neurons in the affective system encode distinct valences (fig. 4.2A). That is, one class of neurons only carries the positive signal of reward, and another class carries the negative signal of aversion. These *valence neurons* code their + or – signal both in their electrical firing patterns and in the kinds of chemicals they release to signal other neurons (such as dopamine, which signals an anticipated reward). Valence neurons have distinctive locations, targets, and networks within reward and aversion pathways. In some places, they form "valence maps" where, for example, + neurons produce pleasure "hot spots" in the brain (fig. 4.2B).[17]

A recent study by William E. Allen and colleagues on mice found another example of particular neurons that encode valences or emotions.[18] This study confirms the close link between the interoceptive and affective types of consciousness (chap. 2), here involving "thirst neurons" in the brain's hypothalamus. Thirst affects, along with food hunger and air hunger, are considered to be "primordial emotions" that control basic drives to maintain homeostasis in the body.[19]

In another new study on mice, Carlos A. Campos and colleagues found a kind of valence neuron that carries "danger" signals—signals from skin pain, loud noises, or a dangerously full stomach. This neuron type is in the parabrachial nucleus of the interoceptive pathway (fig. 2.3), and its signals elicit fear responses from affective brain centers like the amygdala.[20]

With both thirst neurons and danger neurons, it seems that neurons can code affective information that is far more sophisticated than just a simple liking or disliking.

Although the affective system of the vertebrate brain seems diffuse, with its different intercommunicating parts overlapping one another in function, the parts do specialize for their own affective functions. These parts are shown in figures 4.1 and 4.2, and they include the amygdala for fear and emotional learning, lateral habenular nucleus for coding punishment and disappointment, nucleus accumbens for motivated seeking of rewards, a part in the anterior reticular formation for arousal,[21] and the parts of the cerebral cortex that use reason to control one's core emotions and passions (e.g., the orbitofrontal cortex in fig. 4.2). Upon further study, this affective system could prove to have a well-organized and highly differentiated structure, after all, an organization that researchers are just starting to recognize.

Another relatively new finding is that the affective regions relate closely to the premotor brain regions that choose behavioral actions. In mammals, we observe this phenomenon in a heavy input of affective signals to the nucleus accumbens (fig. 4.2), which is a motivation center in the basal ganglia (the premotor region for body movements), and to the hypothalamus (the premotor region for many inner-body actions such as the gut movements of digestion and the heartbeat). As another example, the affective region called the periaqueductal gray signals the motor panic actions of fleeing, curling one's body into a ball for protection, sweating, and so on (fig. 4.2A).[22]

This close relation of valences to motor output is even seen in the primitive behavioral circuits in brains of the invertebrate sea slugs and snails (fig. 4.3; plate 5).[23] These circuits may be too simple to produce true affects, but they still show the principle. Here the incoming sensory signals are assigned positive or negative valences in an "integrator circuit for incentives," which communicates with nearby premotor circuits called central pattern generators (CPG) that control the slug's repetitive feeding and locomotor movements.[24] The valences thus influence whether these two movements are performed or not performed. In the slugs, each of the circuits has only a few neurons, and they are not connected into multiple levels of processors. This simple system provides a distant mirror to that of the vertebrates, whose incentive circuits and premotor circuits have become much more complex, with many levels and subcenters.

So far, our analysis indicates that all the vertebrates have affective as well as image-based consciousness. Affective consciousness differs from image-based consciousness in the neuronal circuits and functions that produce it: for example, in its valence-coding neurons, its many neuromodulator chemicals, and its nontopographical organization.

These two types of consciousness also share similarities, such as hierarchies of pathways, inputs received from many different senses, and centers for attention and memory. And in comparing exteroceptive, affective, and interoceptive consciousness in vertebrates, we find that these diverse conscious states trace not only to the corticothalamic system of mammals but also to the optic tectum of fish and amphibians (for most of the distance

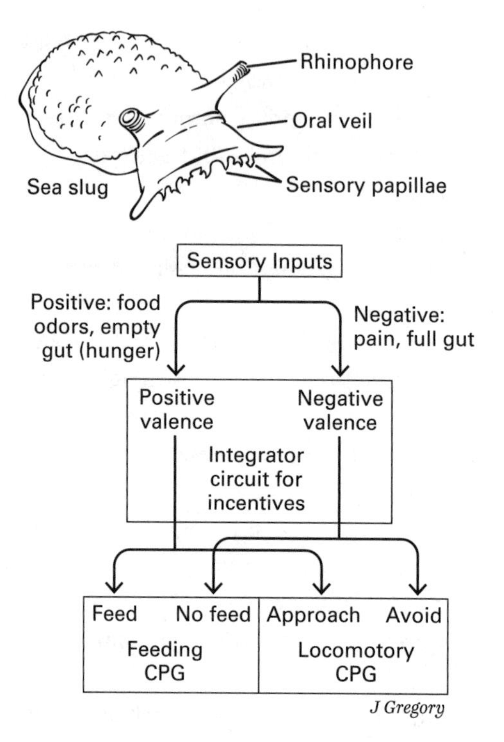

Figure 4.3

Basic valence circuit in the sea slug (*Pleurobranchaea californica*), which may show the simplest ancestral type. *Top*: the slug senses food and other outside stimuli with its sensitive oral veil and also senses the emptiness or fullness of its gut. *Middle*: these stimuli are assigned positive or negative valence in the integrator circuit for incentives, which tells the animal's central pattern generators (CPG) to start or stop the feeding and locomotory movements (*bottom*). See plate 5.

senses) and to the subcortical affective structures of all vertebrates (for most affects).

Therefore, at this point we can deduce several things about the nature of consciousness and subjective experience. First, consciousness in vertebrates comes in various functional and neuroanatomical forms, varying both within brains (exteroceptive versus interoceptive versus affective regions) and across species (e.g., in the tectum of fish versus mammals' cortex). Nonetheless, we can identify some common factors across the types of consciousness and across species. To tell if diversity and commonality are a general feature of all animals that reach a certain level of neural-hierarchical complexity, let us go beyond the vertebrates and see if we can find consciousness among the many and varied invertebrates, and if so, how is it created.

5 The Question of Invertebrate Consciousness

We have argued that the discovery of diversity both within and across species, and the search for the common factors that create consciousness, are central to our naturalized theory of consciousness. So far we have considered the diversity and commonality within the vertebrates. But now an increasing number of investigators are arguing that certain *invertebrates* have consciousness.[1] This proposition is of great interest for a naturalized theory of consciousness, because if some invertebrates are conscious, then that finding greatly expands our sense of what neuroanatomical underpinnings can create consciousness, and provides more data to compare across species and across brain anatomies in our search for common factors.

We have already deduced that the tunicates and amphioxus, the closest invertebrate relatives of vertebrates, are not conscious because their simpler brains and lack of distance senses do not meet the criteria (chap. 3). All the other invertebrates are more distantly related to vertebrates, and most of them are classified as *protostomes*. The protostome group consists of many kinds of worms, including roundworms, flatworms, and earthworms; mollusks; arthropods; brachiopods (lamp shells); and others. All protostomes have a nervous system, and many have a brain.

Among the protostomes, the jointed-limbed arthropods (such as insects, spiders, crabs, and centipedes) and the mollusk group named cephalopods (such as octopuses, cuttlefish, and squid) have the most complex brains, senses, and behaviors, so we have tested these protostomes for consciousness.

We applied the criteria for recognizing consciousness in vertebrates (from chaps. 3 and 4) to arthropods and cephalopods. Let us consider affective consciousness first. Recall that the behaviors that indicate affective experiences are global operant learning, trade-offs, frustration, self-delivery of drugs or rewards, and approaching drugs and conditioned place preference (box 4.1). Arthropods, especially insects, have been shown to do all these things.[2] Cephalopods, despite being quirky and sometimes uncooperative experimental subjects, have successfully met all the criteria for which they have been studied.[3]

Now we turn to image-based consciousness. Recall the vertebrate-derived criteria for recognizing this phenomenon: elaborate sensory organs for the distance senses of vision, hearing, smell, and others; complex chains of neurons in sensory pathways; mapped neural representations in the brain; circuits for attention and memory; and so on. Insects have all these features, including image-forming compound eyes, but the small size of their brains (fig. 5.1; plate 6) raises concern about whether they have enough neurons for consciousness processing.[4] Insect brains have from one hundred thousand to one million neurons, whereas the brains of various vertebrates have from ten million to eighty-five billion neurons.[5] In the case of cephalopods (fig. 5.1), their brains fit all the neural criteria for image-based consciousness that we listed, although (1) the brain circuitry is understudied, and (2) much of their nervous system and many of their behavioral programs are located not in their brains but

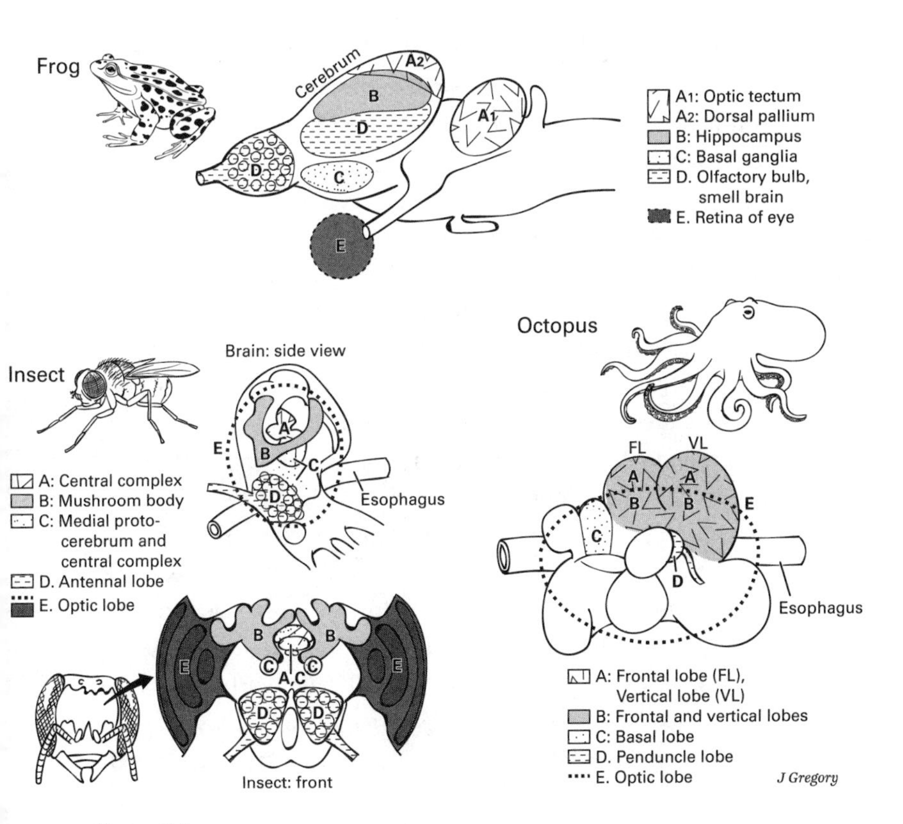

Figure 5.1

Brains of an insect and an octopus compared to brain of a vertebrate (frog). Mostly shown in side view. Regions with similar functions for consciousness are colored similarly in the three kinds of brains. For more information, see Barron and Klein 2016; Shigeno 2017. Despite the functional similarities, the three brains evolved independently of one another from the simple nervous system of a brainless ancestral worm (chap. 7). The general code: A, image-based consciousness; B, memory; C, premotor center; D, smell processing; E (with the dotted outlines), visual processing. See plate 6.

out in their arms. Still, the complex brains of cephalopods are the largest among the invertebrates, containing about fifty million neurons.[6] The cephalopod brain is as large, relative to body size, as that of a reptile.

When we considered some of the protostome invertebrates with less-complex brains—roundworms and flatworms—we found they did not meet most of our criteria for consciousness. The well-studied roundworms (nematodes) evidently lack affective consciousness; that is, they show only classical learning with little evidence for global operant learning and none for unlimited associative learning (chap. 4).[7] When we turn to image-based consciousness, roundworms fail again. They can integrate sensory information from their simple touch-, chemical-, and light-detecting senses, but they have no elaborate distance senses by which to model the space around them; and their memory is limited.[8] Thus roundworms have no blueprint of where they are. When they lose a sensory trail as they search for food or mates, they search systematically but blindly to try to pick up the trail again. This is different from conscious animals, whose directed foraging indicates they have a spatial plan in mind.[9]

Excepting cephalopods, most mollusks have little or no brain. However, the gastropod mollusks—snails, slugs, sea hares, and their kin—have a set of brain-like ganglia in their head. The particular gastropods that are studied most often, namely, the sea hare *Aplysia* and the sea slug *Pleurobranchaea* (figs. 4.3 and 6.3), have simple ganglia and sensory organs and show only simple, nonglobal types of operant learning, so it is doubtful whether they have primary consciousness. However, some gastropods have image-forming eyes (e.g., the great land snail *Lymnaea*) or are capable of "second-order learning" that approaches the learning criterion of Bronfman and colleagues (the land snail *Helix*), so these species may be on the cusp of consciousness.[10]

So far, arthropods and cephalopod mollusks are the invertebrates most likely to have consciousness. Cephalopods are currently the darlings of the animal consciousness field, and two fine books have recently summarized their remarkable intellectual abilities and the evidence that they are conscious: Sy Montgomery's *The Soul of an Octopus* and Peter Godfrey-Smith's *Other Minds: The Octopus, the Sea, and the Deep Origins of Consciousness*.[11] We also mention the many studies of Jennifer Mather, Binyamin Hochner, and others.[12] These books and articles are recommended, and here we will just repeat the anecdotal support for octopuses having consciousness. For instance, evidence suggests that octopuses can recognize individual humans (image-based consciousness), to whom they take a liking or a dislike, squirting the disliked people with water from their siphons (affective consciousness). They may also play, with plastic bottles in a jet of water in the Seattle Aquarium, for example; and some investigators posit that play indicates consciousness because it can show the "joy" of positive feelings.[13]

Consciousness in arthropods, on the other hand, is much more controversial given their small brains.[14] Some of the controversies are over whether some arthropods are conscious and others are not, and whether all arthropod species must have the same kind of consciousness.[15] We think that the evidence favors the view that all the classes of arthropods have consciousness, because their brains share all the same parts. In fact, the fossil evidence indicates that the arthropods have shared their basic brain structures since the very first arthropods appeared over a half billion years ago.[16] For our purposes, however, showing that just one arthropod is conscious would be enough to tell us about the diversity and commonalities of consciousness in the animal kingdom.

To this end, we will describe two especially revealing studies of the behavior of bees, whose brains are among the largest, and whose behaviors among the most complex, of all insects. One study indicates that bees have image-based consciousness, the other that they have affective consciousness (figs. 5.2 and 5.3). The image study was conducted by Karine Fauria, Mathew Colborn, and Thomas Collett.[17] In their experimental design, a bee flew within a long box to reach a hole at the far end that led to food (fig. 5.2). The bee had been trained to know which (of

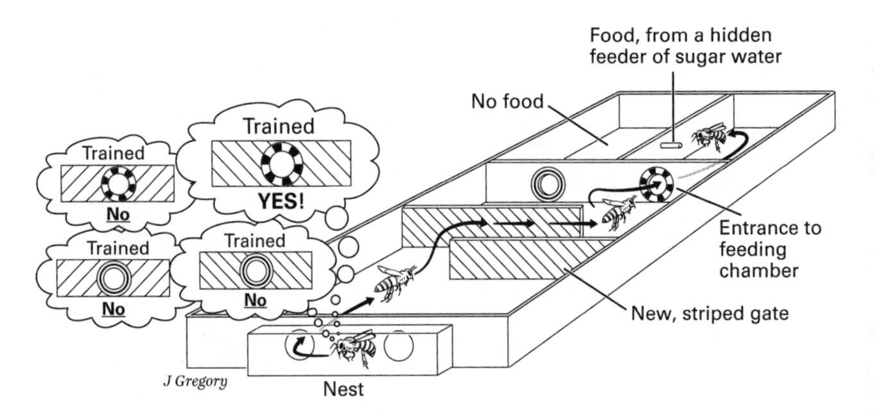

Figure 5.2
Test of whether bees have image-based consciousness. Bee travels from its "nest" chamber through a long box, seeking hidden food at the far end. Thought balloons at left show that the bee has already been trained to know which one of four combinations of stripe-and-bull's-eye patterns leads it to the food. Now, however, the striping and bull's-eye are not presented together, as in training, but the striping instead appears on a gate that blocks the view of the bull's-eyes. To reach the food, the bee must consciously remember the correct stripe-and-bull's-eye patterns. From Fauria, Colborn, and Collett 2000.

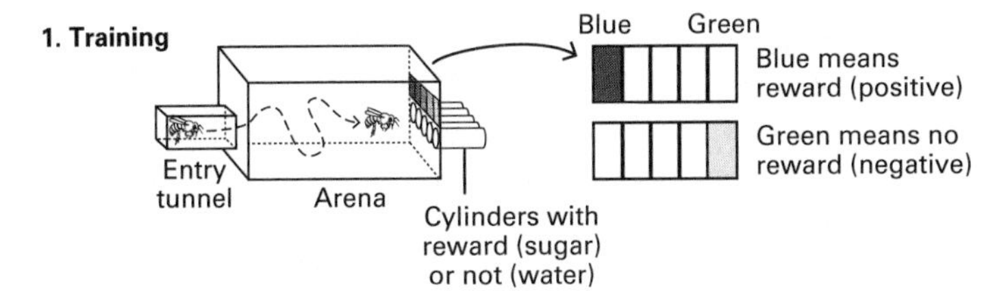

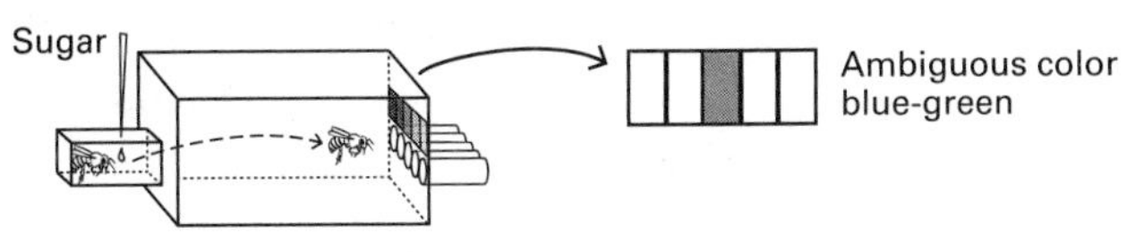

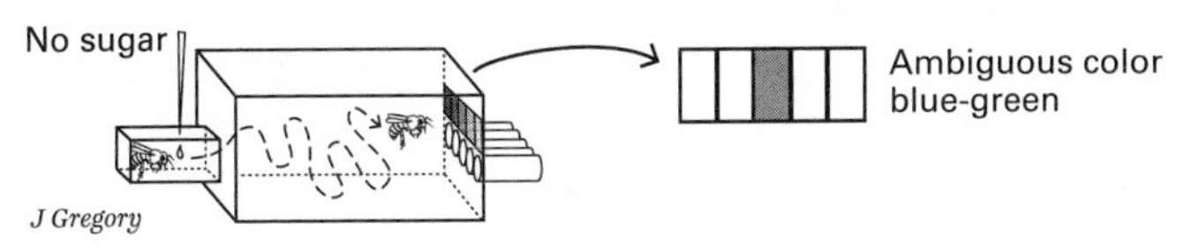

Figure 5.3

Test of whether bees have affective consciousness: the judgment bias test. 1. Bee is trained to know that a blue flag marks the entrance to a cylinder with a big sugar reward, but a green flag marks the entrance to a cylinder with plain water (no reward). 2. In the test, the bee is given a small taste of sugar in the entry tunnel and then shown a flag of intermediate, ambiguous color. 3. For the "control" part of the test, the bee gets no sugar taste before it goes to encounter the ambiguously colored flag. Overall, the test measures if the bee primed with the sugar taste enters the ambiguously marked cylinder more quickly, which would mean the priming influenced the bee positively. Bees passed the test. Simplified from Perry, Baciadonna, and Chittka 2016.

several) combinations of a striped-plus-bull's-eye pattern around the hole signified that food was behind it. But in the real test, the task was made more difficult, and the bee could succeed only if it had recorded and recalled images of *both* the patterns. The bees succeeded, meaning they had formed visual mental images. Thus bees have image-based consciousness.[18]

The study that tested if bees have affective states was performed by Clint Perry, Luigi Baciadonna, and Lars Chittka (fig. 5.3).[19] The authors used the judgment bias paradigm, a well-established test for showing if emotions exist in humans and mammals, though it had never before been applied to any non-mammalian animals.[20] Its premise is that "subjects in a positive emotion state tend to respond to ambiguous (intermediate) stimuli as though predicting the positive event." The procedure first trained bees to know that one color signified a big food reward (sugar) and another color signified no reward; then it tested how the bees responded to an *intermediate* color, either if they had been primed first with a tiny sugar taste or if they had not. Results showed that, indeed, priming with sugar made the bees more likely to respond to the intermediate color when looking for food. The bees passed the judgment bias test, and the authors cautiously concluded that bees have "emotion-like" state changes. To us, however, this indicates bees do experience positive affects, or even a few moments of elevated good mood. With this finding, we can add bee consciousness to cephalopod consciousness and vertebrate consciousness.

In conclusion, while most of the invertebrates lack consciousness, arthropods and cephalopods have it, as do the vertebrates. Vertebrates are only distantly related to protostome animals, and arthropods and mollusks are distantly related to each other within the protostomes. Therefore the brains and consciousnesses of the three groups must have evolved independently. Biologists call this phenomenon "convergent evolution of the same characteristic," and a philosophical term that applies is "multiple realizability" (many different ways to build the same thing).[21]

Vertebrates, arthropods, and cephalopods include the most active animals on the planet. Each moves through a complex environmental space, and they have the most complex sets of sensory organs. This is why we say consciousness also correlates with good locomotory and navigation abilities.

The findings about invertebrates add even more support for the diversity of consciousness. It appears that any animals that possess elaborate distance senses and a sufficiently complex, hierarchically organized brain—a set that includes all vertebrates, cephalopods, and arthropods—regardless of their evolutionary lineage, are capable of consciousness. But this means we are also finding the shared features of these diverse conscious brains. What do tallying and formalizing the commonalities tell us about the universal principles that create consciousness? We address that question in the next chapter.

6 Creating Consciousness: The General and Special Features

We have identified several different forms of primary consciousness (image-based exteroceptive, interoceptive, and affective) that are diverse in that they differ in their neural bases and in the qualitative experiences they create (chap. 2). Also, while some scholars claim that consciousness is confined to humans, mammals, or mammals and birds, and that the so-called neural correlates of consciousness require an advanced cerebral cortex and thalamus (and the equivalent regions of bird brains),[1] we have found instead that the neural substrate of consciousness is diverse, and all the vertebrates and some invertebrates have consciousness (chaps. 3–5). Therefore the neural basis of conscious images and affects need not include the cerebral cortex, but less-elaborated brain structures can suffice.[2]

If consciousness is so diverse both within individual brains and across species, we want to identify the common features that could naturally create consciousness and would at the same time allow for and help explain the unique features of subjectivity. We started to recognize such shared ingredients in chapters 3 through 5, and we assemble and catalog them in this chapter (boxes 6.1–6.3). Later, in chapter 8, we will use these common features to address the natural creation of subjectivity and the explanatory gaps.

Box 6.1

The defining features of consciousness, Level 1: General biological features, which apply to all living things

Life, embodiment, and process

- Life: use of energy to sustain self, growth, responsiveness, reproduction, and adaptiveness to change. All known life is cellular.
- Embodiment: a body with an interior separated from the exterior world by a boundary.
- Process: life *functions* are complex, dynamic processes, not material things.

System and self-organization

- System: entity considered as a whole, in which arrangements and interactions between the parts are important.
- Self-organization: interactions of the parts organize the patterns at global level of the whole system.

Hierarchies

- Hierarchy: complex system with different interacting levels, organized from simpler to more complex. Levels may be nested within one another: e.g., macromolecules to cells to organs to the organism. New emergent features may naturally appear in the whole system by the addition of new levels and their interactions with lower levels.

Teleonomy and adaptation

- Teleonomy: biological structures perform programmed, goal-directed functions.
- Adaptation: a teleonomic structure or function as evolved by natural selection.

Box 6.2

The defining features of consciousness, Level 2: Neuronal reflexes and simple core brains

Rates

- Reflexes and all neural communications are extremely fast relative to other large-scale physiological processes. Thus they can move a large body in response to a stimulus.

Connectivity

- Simple reflex arcs are chains of several neurons connected at synapses. More complex arcs have more neurons in the chain (C) and in networks (N); they also have more sensory input (S), more neuronal interactions, (I) and capacity to process more information (P).

Increasing complexity

- Further increase in CNSIP leads to complex nervous systems and consciousness.

Basic motor programs from central pattern generators

- These nonconsciously control essential, repetitive behaviors.

Core brain features

- Modulatory, sensorimotor-integrating centers for arousal and directing attention.
- Complex reflexes for inner-body homeostasis.
- Rhythmic locomotion and other basic motor programs.
- Automatic, not conscious.

Box 6.3

The defining features of consciousness, Level 3: Special neurobiological features, which apply to animals with primary consciousness

Neural complexity (more than in simpler, core brain)

- A brain with many neurons.
- Many subtypes of neurons.

Elaborated sensory organs

- Image-forming eyes; multiple mechanoreceptors for touch; separate chemoreceptors for smell and taste.
- High locomotory mobility, to gather the abundant sensory information.

Neural hierarchies with unique neural–neural interactions

- Extensive reciprocal (reentrant, recurrent) communication occurs within and between the hierarchies for the different senses.
- Synchronized communication by brain-wave oscillations may be required for sensory binding and generating mental images.
- Higher levels of the hierarchy allow the complex processing and the unity of consciousness.
- Hierarchies allow consciousness to predict events a fraction of a second ahead in time.

Neural pathways that create mapped mental images or affective states

- Isomorphic representations: neurons are arranged in topographic maps of the world or body structures.
- Affective states: from valence coding rather than from topographic mapping; the hierarchy and networks are more diffuse with more centers and more use of neuromodulators.
- Both feed into premotor brain regions to motivate and guide behaviors in space.

Box 6.3 (continued)

Attention as a participant in consciousness

- Selective attention mechanisms in brain: for focusing consciousness onto salient objects in the environment. Related feature of *arousal* is also present, adjusting the level of consciousness.

Memory

- Short-term, minimal sensory memory is needed for the continuity of experience in time. Longer-term, higher-capacity memory evolved soon after consciousness arose. However, we cannot rule out more memory being present from the start, before consciousness arose.

In this chapter, we first identify the most basic ingredients in the creation of consciousness as the *general biological features* of all life (box 6.1), from which consciousness evolved and whose properties are essential to consciousness and subjectivity.

Second, to the general features of life are added neurons, which are assembled into reflexes, networks, and then simple brains, which regulate basic survival behaviors (box 6.2). These characteristics, while not conscious themselves, are indispensable to the creation of primary consciousness and subjective experience.

Third, to the reflexes and core brain are added the unique *special neurobiological features* that enable all the diverse forms of consciousness (box 6.3). These features are unique to consciousness and not found in animals that lack consciousness. These are the features that we believe allow brains to cross the threshold into consciousness. We attempt to show that these three levels of features accrued naturally and continuously without requiring any new properties of the *physics* of brains or any "mysterious" properties along the way.

General Biological Features (Box 6.1)

As discussed in chapter 1, all life is characterized by many unique principles and functions that are not found in inanimate nature and contribute to the creation and evolution of consciousness and subjectivity.

For instance, all living things, even single-celled animals and plants, are *embodied systems* with an interior that is separated from the external environment. This is a general biological characteristic that endures as a critical feature of consciousness. Both life in general and consciousness in particular are features of embodied animals and rely on a bounded, material body or brain, respectively, for their existence.

Since both life and consciousness occur only in embodied organisms, then all the features of both life and consciousness are personal and unique to the living individual. So in this sense, the personal nature of consciousness is already rooted in the personal nature of life.[3] This point is essential if we are to understand the nature of subjectivity. We return to it in chapter 8.

Second, life and consciousness are not structures or concrete things but rather *processes*. In biology we call the processes of life "physiology." In consciousness research, the terms "consciousness," "mind," and "subjective experience" actually refer to processes that the brain performs. In this sense, even using the nouns "primary consciousness" and "qualia" is slightly misleading, since what we are really talking about are processes (verbs), not things (nouns).[4] In any event, we should keep in mind that consciousness is an aspect of what the living brain *does*. It follows that any theory of the neural correlates of consciousness must be a *functional* explanation based on neuroanatomical features.

Third, as emphasized by Mayr and many others, each living thing is a system made of organized parts, and the unique functions of living systems depend on their *hierarchical* structures.[5] Mayr called the biological hierarchies, whose levels resemble a series of boxes nested within each other, from small to large, *constitutive hierarchies*. For example, the series comprising atoms, molecules, macromolecules, organelles, cells, tissues, organ is a constitutive hierarchy (fig. 6.1).

One interesting feature of constitutive hierarchies is that when higher, more complex levels are added to and interact with the lower levels, the system as a whole acquires new or even novel (never-before-existing) features. These new features are sometimes called "emergent properties."[6] For example, an atom or molecule within a body is not "alive" in and of itself, but the property of "life" is a nonmysterious emergent system property of cells and larger organisms.

Much has been written about "emergence" with reference to consciousness,[7] most of which is beyond the scope of this book. Our view, in agreement with Searle,[8] is that emergence applies to brains and consciousness just as it applies to the rest of nature. However, brains and consciousness involve more—and more complex—levels than elsewhere in biology, with more kinds of uniquely new features.

Other general features of life are *teleonomy* and *adaptation*. These terms simply mean that living organisms perform programmed, goal-directed functions, most of which are dictated by the genes. An adaptation is a goal-directed function (or a goal-meeting structure) as evolved by natural selection but not by predesign and is not necessarily directed by a conscious will.[9] Living things and processes do have adaptive *purposes* in this

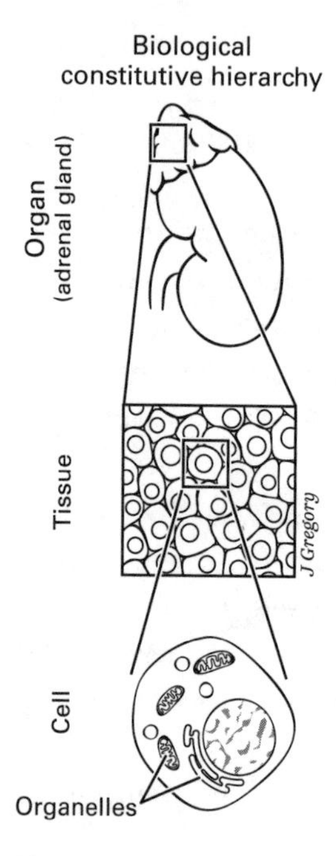

Figure 6.1
Constitutive hierarchy: a biological hierarchical system in which the higher levels consist of the lower levels. The adrenal gland—an organ above the kidney that secretes chemicals to help the body cope with stress—consists of tissues that consist of cells that consist of cellular organelles. The organelles in turn consist of macromolecules, then smaller molecules, then atoms (not shown).

sense. Later we will discuss how neural reflexes and sensory consciousness, like many other biological processes, are adaptive and beneficial to survival.

So we have learned that all living systems have many features that are essential to the subsequent creation of consciousness and subjectivity. This is important to our theory of neurobiological naturalism, because while we recognize consciousness and subjectivity as truly unique in nature, we see that they cannot be explained or understood without considering them in the context of the rest of biology. This, of course, was Searle's view, as well, and why he called his theory "biological naturalism."

The Level of Neurons, Reflexes, and Simple Core Brains (Box 6.2)

Neurons and Reflexes

While even single-celled organisms have the general life features, the next step on the road to consciousness requires a multicellular animal body in which different kinds of cells have different functions, including the neurons that perform rapid communication. As described earlier, the simplest arrangement of neurons is a reflex arc made of a few neurons: sensory, motor, and, usually, interneurons (fig. 2.1B). Neurons carry signals rapidly and communicate quickly at synapses, so that neural processing is fast. Although reflexes do not involve consciousness, and the animals that rely only on reflexes are not conscious, reflex arcs are the building blocks of the neural substrate that makes consciousness possible.

Recall that reflexes are fast, automatic responses to stimuli from outside or inside the body, as signaled by reflex arcs. In the vertebrates, most reflexes involve the spinal cord and brain stem, but in fact, they require only basic neuronal connections, as in a nerve net that spans the body surface of a brainless jellyfish.[10]

While reflexes come from nerve cells and therefore have all the general features of cellular life, they also have something more: the unique additional ability to signal widely and quickly across the large bodies of multicellular animals. This coordinates distant parts of the body to act together appropriately.

Our framework roughly divides reflex arcs and reflexes into simple and complex forms. The simple is expanded into the complex by adding more interneurons and more cross talk between these additional interneurons, for sensory integration and for commanding body movements. Figure 6.2 shows how expanded reflex arcs signal intricate forward-and-reverse escape

Figure 6.2

Expanded reflex arcs allow complex reactive behaviors and represent an early step toward consciousness. These arcs are more complex than the one shown in fig. 2.1B. These expanded arcs, with several orders of interneurons (*hexagons*) between the sensory neurons (*triangles*) and motor neurons (*circles*), control escape movements in the roundworm *Caenorhabditis elegans*. The worm, pictured at top, moves forward or backward by bending its body into traveling waves. Plus and minus signs indicate that the axon of the neuron either stimulates (+) or inhibits (–) the activity of the next neuron. By tracing the circuit, you can see that touch to the anterior body (*left*) leads to backward locomotion, so that the worm withdraws from the threatening stimulus in front of it (path **abcd**). This also inhibits all forward locomotion toward the stimulus. On the other hand, touch to the posterior body (path **efgh**) causes forward locomotion while inhibiting backward locomotion, so that the worm moves away from the threatening stimulus behind it. (Specific neurons are labeled with the standard abbreviations for their names, AM, PM, etc., the meanings of which are not important here.) From the studies of Mark Alkema and colleagues.

movements in a roundworm.[11] From such arcs, more levels of sensory-processing interneurons evolved, like additional links in a chain, forming a multilevel hierarchy. Even the most complex reflexes are not conscious, but elaboration of reflexes was the royal road to consciousness in the first vertebrates and arthropods, as well as in cephalopods.

In the more advanced nonconscious animals that have attained this reflexive stage, the central nervous system contains basic motor programs from central pattern generators (fig. 6.3).

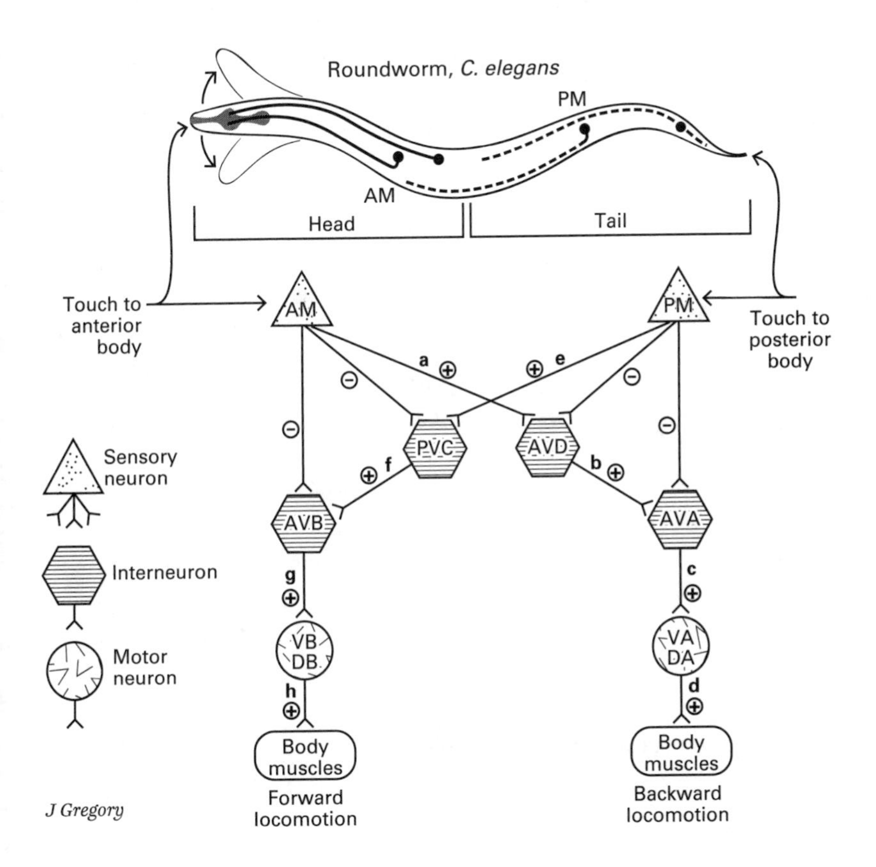

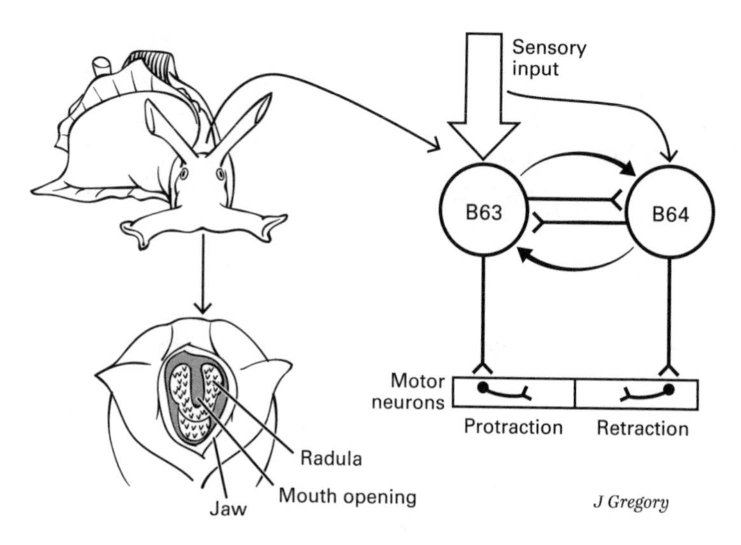

Figure 6.3
The central pattern generator (CPG) that controls feeding cycles in the snail-like sea hare, *Aplysia*. It pushes out and pulls back a scraping-biting structure in the mouth called a radula (shown below the animal). The CPG is based on two brain neurons named B63 and B64. First, sensory neurons detect the food (at the big white arrow) and signal B63, which starts a feeding cycle by protracting the radula and opening the mouth. Then B63 tells B64 to signal retraction. These two neurons send signals back and forth, to continue the protraction–retraction cycles of feeding. For a more complete explanation, see Jing et al. 2004.

These CPGs fire rhythmically to signal repetitive actions such as the persistent mouth movements by which sea hares take in food. Also recall the locomotion and feeding CPGs of sea slugs (fig. 4.3; plate 5).

Core Brains (Still Not Conscious)

In multicellular animals that move through their environment, sensory receptors are most abundant in the head end because that is the first part of the animal to encounter new stimuli to sense. Correspondingly, more sensory processing occurs in the head region of the nervous system, so that is where brains evolved. This primitive brain is where the original reflex arcs elaborated the most, into neural circuits for basic survival functions. These core brain functions are not conscious, but they are the next step on the road to consciousness.

In vertebrates, the brain region involved is located in the brain stem and some of the diencephalon (shaded area in fig. 6.4). In arthropods, this region includes the "protocerebrum" (the circled blue Cs in the insect brain in plate 6).[12] We do not know enough to locate this core in the brain of cephalopod mollusks.

The core brain receives all kinds of sensory inputs. In vertebrates, it organizes these inputs to adjust homeostatic functions, that is, to control respiratory, heart, and digestive functions. To this end, the core brain is also involved in many complex and unconscious reflexes: the reflexes for swallowing, vomiting,

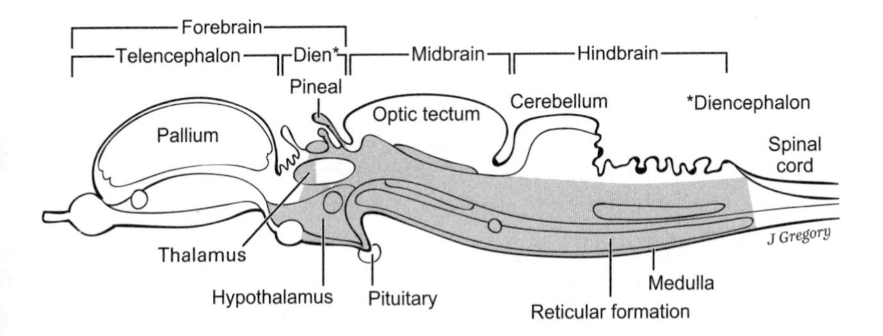

Figure 6.4
Core region of the vertebrate brain for basic survival functions (*gray shading*). This core originally was not conscious, in our view. Most of brain stem and part of diencephalon. Based on fig. 4.1.

gasping, coughing, and keeping the body's blood pressure from getting too high or too low. The core brain contains many of the CPGs for the basic rhythmic motor programs, such as breathing patterns, and for the rhythmic body movements of the swimming fish or the walking mammal. Most important for the evolution of consciousness, the core controls whatever levels of arousal and motivation the animal may show, and the animal's ability to direct its attention to certain stimuli and ignore other stimuli. We assume that the core has performed these arousal-attention functions since its ancestral, nonconscious days.[13]

In chapter 3, we introduced the *reticular formation* of the vertebrate brain as important for arousal and attention, and it belongs to the core (see also fig. 4.1). Its neurons not only receive inputs from, but also project their axonal outputs to, many other parts of the brain, meaning that the reticular formation can arouse most of the brain at once.[14] The invertebrate cousin of the vertebrates, the entirely nonconscious amphioxus larva, has an equivalent region in its simpler brain, a "core integrative center." Seen in figure 3.6, it is where all the types of sensory neurons are projecting to the "motor center," which then projects to many motor neurons. Lacalli interprets this hub region as a motivational center that receives lots of sensory input and then signals the larva's basic swimming responses.[15] In the process, it determines how sensitive the larva will be to the incoming stimuli and how readily the larva will respond. Thus this center *modulates* the sensory input to determine how motivated and how easily aroused the amphioxus will be. Motivation and arousal—these are basic, nonconscious functions of the core of the amphioxus brain that match the same functions in the core of the conscious vertebrate brain.

Arthropods (insects) also have brain neurons for arousal and directing attention, though these evolved independently from those of vertebrates. They include dopamine-releasing neurons that distribute widely, in clusters, throughout the fruit fly brain.[16] With such a broad distribution, they produce widespread arousal of the entire brain (as does the reticular formation of vertebrates). As in the vertebrates, these neurons also influence the fruit fly's locomotor activity.[17] To this we should add that the core protocerebrum of the insect brain (fig. 5.1; plate 6) modulates sensory inputs and adjusts motor responses based on the *motivational state* of the animals.[18] Again, this matches what the core of the vertebrate brain does.

Therefore the ancient animals that reached the cusp of consciousness—of both the prevertebrate and the prearthropod lines—had a core brain that controlled attention and arousal, as well as body homeostasis, essential locomotion, and basic survival behaviors. But these animals still did not have consciousness. They were not aware of what their receptors sensed, did not *feel* any affects toward the things they approached or avoided.

Because of this lack, the behaviors of animals at this stage of reflexes, basic motor programs, and core brain functions are limited. They are at the stage of the roundworms mentioned in the previous chapter, who follow stimulus trails in search of food and mates but get lost when they lose such a trail because they have no mental map of space.[19]

The Special Neurobiological Features of Consciousness (Box 6.3)

We now turn to the neural features that are greatly enhanced in conscious nervous systems but are absent (or much simpler) in animals that do not possess consciousness. Although these special features derive from normal evolutionary processes, they are the keys to understanding consciousness and subjectivity.[20]

These are the features we began to identify in chapters 3 through 5, in both vertebrates and invertebrates, as indicating image-based consciousness and the states of affective feeling. We list them formally here and explain some of them more thoroughly.

Neural complexity, including a complex brain. All three groups of animals that we identified as conscious have complex brains with a minimum of about 100,000 neurons. Their brains also contain many different subtypes of neurons, which have distinct shapes and roles.[21]

Elaborated sensory organs. Unless one can detect many different kinds of stimuli in detail, it is impossible to form a sensory image of the world or of one's body or to aim one's responses precisely at a target when using affect-driven behavior. Detection of this sort requires eyes that form a detailed image of visual space, a nose sensitive to many odors, ears for hearing and balance, taste receptors, and proprioceptors to sense one's own movements. These different senses are called different *modalities*. Vertebrates, arthropods, and cephalopods have all these complex senses, and the water-dwelling members of these groups have additional receptors for sensing the vibrations caused by other animals moving through the water.[22]

Neural hierarchies and neural integration with unique neural-neural interactions. Neural hierarchies differ from the constitutive hierarchies of biology (see box 6.1) in the way their higher and lower levels are physically related. While the lower levels of the cells and tissues of an organ like a kidney are contained within successively higher levels of the hierarchy (fig. 6.1), the lower levels of neural hierarchies need not be *physically contained* with higher levels. For example, in vertebrates, structures in the spinal cord are not physically contained within the lower brain stem, and the lower brain stem is not physically contained within the tectum (see fig. 3.1 and plate 1).

However, the amount of interconnection is extreme. Synaptically connected chains and hierarchies of rapidly communicating neurons in the more complex neural networks allow multiple levels of neural processing to interact intimately without lower levels being physically nested within higher levels. This interconnection creates an exponential increase in the back-and-forth communication among neurons, both within and between levels of the neural hierarchy, allowing an array of new features that we find only in complex neural hierarchies.[23]

The official name for such feedback and cross talk among neurons is *reciprocal, reentrant,* or *recurrent communication.*[24] It also occurs between the different hierarchies for the different senses, for example, so that the hearing hierarchy informs the visual hierarchy. Overall, the reciprocal communications are coordinated, possibly by synchronized, back-and-forth electrical oscillations that run through the neural networks (brain waves). This synchronized communication is thought to bind all the information from the different senses together into a unified image or experience.[25]

Let us emphasize the integration aspect. Giulio Tononi's integrated information theory of consciousness proposes that the integration and unification of subparts are critical for how the brain creates consciousness.[26] For image-based consciousness, the actions of the different levels and parts of the exteroceptive pathways must be integrated—or else they could not construct a unified topographic map of the world from all the different senses (chap. 3). For affective or emotional consciousness, though its contributing parts are less rigidly hierarchical and not topographically organized, their actions must also be highly integrated, as indicated by the extensive cross communication among the brain regions involved in affects (fig. 4.2).

What is the *value*, for consciousness, of having a multilevel neural hierarchy? First, as we have stated several times, the higher levels that were added to the original core brain allow the more complex neural processing that consciousness demands. Second, as we hinted in the previous paragraph, the added levels allow the pathways of the different senses to interact with each other more extensively, not only to generate a unified world image but also to allow motivational unity toward a motor response (i.e., to motivate you to perform just one responsive action at a time). The important centers for the multisensory convergence include the optic tectum of fish and amphibians, the cerebral cortex of mammals, the affective subcortical regions of all vertebrates, the central complex of arthropods, and a part of the brain in cephalopods called the superior lobe (figs. 5.1, 4.1).[27]

From a different perspective, however, consciousness is valuable because it enables *prediction* processes, for which neural hierarchies are essential.[28] The predictive mechanism, as explained in figure 6.5, depends on up-and-down communication within the hierarchy. As long as the predictions are kept updated, they stay a fraction of a second ahead of what is happening in the world—a huge benefit for survival whenever the conscious animal successfully predicts the attack path of a predator (soon enough to avoid predation) or the escape path of a prey animal (soon enough to adjust and catch the prey).

Neural pathways that create mapped mental images or affective states. While image-based consciousness and affective consciousness share the special features of complex, hierarchical, reciprocally communicating networks of neurons, they differ in some ways: mainly in topographic mapping versus valence coding, respectively. This is a good place to reiterate that in the vertebrates the mapped images and affects also involve different (but overlapping) regions of the brain. Figure 6.6 (plate 7) summarizes these two different regions for consciousness in the vertebrate brain. (In arthropods and cephalopods, we do not know the locations of the two conscious subsystems well enough to tell if they involve such distinct brain regions.)

Attention mechanisms in the brain. Directing attention toward a stimulus can occur consciously as well as unconsciously. The brain regions for consciously focusing attention on important stimuli include the optic tectum (chap. 3) and cerebrum of vertebrates, with the help of the reticular formation of the brain stem.[29] In insects, researchers have tied many different parts of the brain to selective attention,[30] so the process may be more widely distributed than in vertebrates. All who study cephalopods agree that they are highly attentive animals, but their brain has not yet been studied for its centers of conscious attention.

Sensory memory. Memory capacity must have increased greatly around the time when consciousness first evolved, in the first vertebrates and the first arthropods. The first piece of evidence for this increase is that learning expanded from the originally

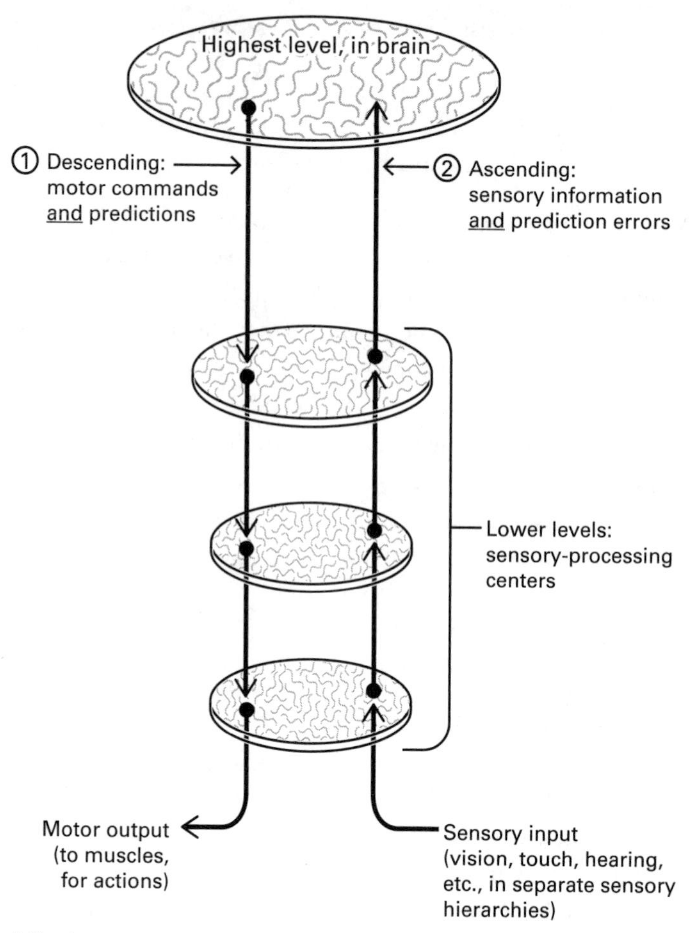

J Gregory

Figure 6.5

Prediction. Neural hierarchies are needed for the role that consciousness plays in predicting events and the best motor responses. (*1*) The highest brain level, using the mapped images of the world it contains, starts to send out motor commands for targeted actions and also sends out its predictions of what these actions will do. The lower, sensory-processing levels receive the predictions and compare them with what the ascending sensory input (*right side*) tells is actually happening in the world and body. The difference between the original prediction and the sensed record is called the *prediction error*, and it is sent up to the highest brain level (*2*). Then the highest level minimizes this prediction error and updates its new action commands and predictions accordingly. The process continues, leading to an ongoing feedback-based adjustment that keeps actions and predictions up to date. As shown, the hierarchy's reciprocal up-and-down communication is the key to the whole process. Squiggle lines in the ovals signify neural processing *within* the levels of the hierarchy.

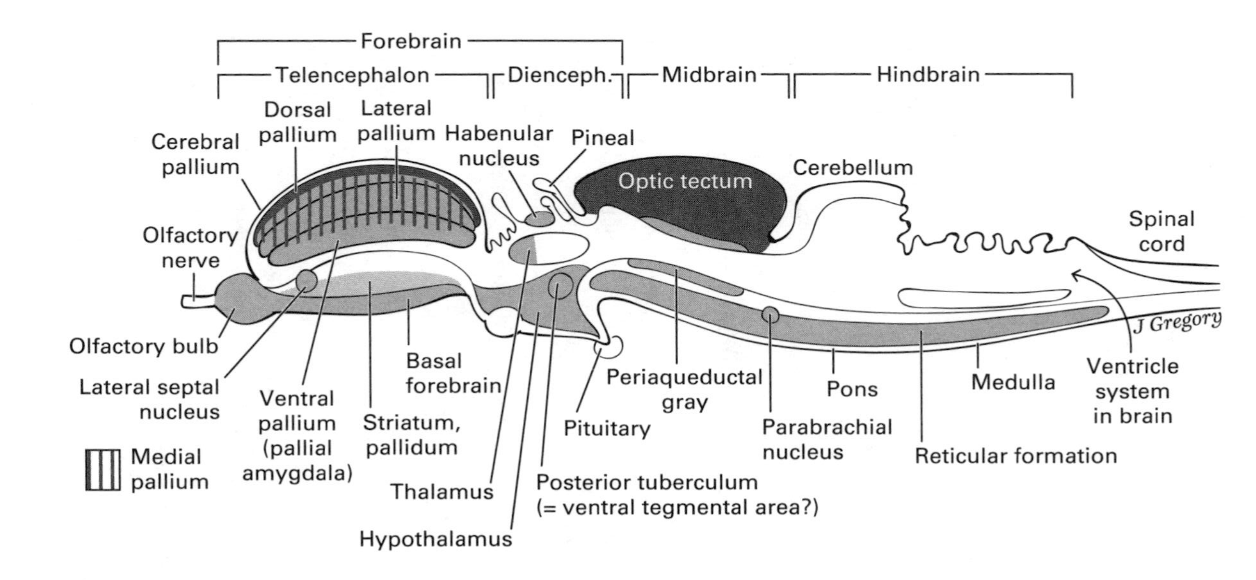

Figure 6.6

Regions of the vertebrate brain for image-based consciousness versus affective consciousness. These regions are mostly different, despite some overlap and much intercommunication. *Lighter gray*, regions for affects; *darker gray*, regions for mapped images. See plate 7.

simple types (sensitization, habituation, and simple associative learning) to global operant learning, as explained in chapters 4 and 5. The second piece of evidence for increased memory is that fish and arthropods react to objects in their complex spatial environments as if identifying the objects from recall.[31]

This finding presents two related problems for understanding the nature of consciousness. First, what role do memory functions play in its creation? And second, did this rapid advance in memory functions in the Cambrian occur before, during, or after sensory consciousness evolved?

Our approach to the first question is to ask: what is the minimal memory requirement for sensory consciousness? Christof Koch conservatively reasoned that image-based consciousness needs at least a fleeting type of information storage called *iconic memory*.[32] Actually, a better name is *sensory memory*, with the term "iconic memory" serving only for visual image perception (lasting less than half a second), "echoic memory" for sound perception (lasting about three seconds), "tactile memory" for touch (lasting about ten seconds), and "odor memory" for smell (lasting longer). The term "sensory memory" includes all these functions.

According to Koch's reasoning, the first conscious animals must have had at a minimum such sensory memory to provide for the continuity of sensory experience in an uninterrupted stream of consciousness. That is, sensory memory of this sort is the least required to construct and maintain a unified sensory experience through time. Koch also suggested that sensory memory functioned to ensure that even briefly sensed stimuli last long enough to trigger a conscious perception.

So this brief type of sensory memory goes among our special features of consciousness in box 6.3. Koch proposed that this kind of memory results from reciprocal communications reverberating up and down a sensory neural hierarchy before fading out. He also said it would entail attentional processes, suggesting that attention and sensory memory coevolved.[33]

Turning to our second question about memory, we hypothesize that soon *after* primary consciousness evolved, the storage capacity and duration of memory greatly increased. The recalled images and affects could then be used to help *interpret* the newly arriving stimuli and the newly generated feelings, thus allowing more advanced forms of learning plus the prediction of anticipated results. Important brain regions for constructing and storing longer, detailed memories are the hippocampus of all vertebrates, the mushroom bodies of insects, and much of the brain above the esophagus in cephalopods (fig. 4.1 and plate 6).

But perhaps we are being too conservative in saying that the first conscious animals had only brief memories with a minimal capacity for storage. Other considerations suggest they had a better memory than this. That is, our vision-first theory says that consciousness evolved to simulate the detailed visual images sensed by newly evolved camera eyes—and considerable memory would already have been needed to record and retain such detailed images from instant to instant. And, starting immediately, there would have been strong natural selection for an increase in memory capacity that allowed long-term recall of a predator, mate, or food item during subsequent reencounters—a recall of the affects that were felt during the initial encounter, as well as of a mental image of how the predator, mate, or food looked. In conclusion, whether conscious memory was originally short or long, memory has always contributed to consciousness and its adaptive value.

Summary of the General Features, Reflexes, and Special Features

As we deduced in chapters 3 and 4, all vertebrates have the numerous general features, reflexes, core brain, and special neural features required for both the image-based and the affective types of consciousness. And as described in chapter 5, arthropods and cephalopods match the vertebrates feature by feature.

Rather than there being any discontinuity or mysterious leap between nonconscious and conscious species of animals, we showed that a stepwise progression, starting with the basic building blocks of life itself and the accretion of novel neural features, created consciousness in a seamless fashion (boxes 6.1–6.3). The general features, reflexes, and core brain evolved into the special features without interruption.

Note that this explanation is entirely natural, based on evolution, life processes, and the unique biological features of complex brains. That is why we call our theory neurobiological naturalism. In the next chapter, we trace the uninterrupted evolution of consciousness through Earth's history, as revealed by the fossil record.

7 The Evolution of Primary Consciousness and the Cambrian Hypothesis

The Cambrian Explosion of Animals

We have argued that consciousness and the mysterious aspects of subjectivity have an entirely natural explanation. More evidence for our hypothesis comes from documenting when and how consciousness evolved step by step so long ago. Now we trace that evolution, and our findings further demystify the nature of subjectivity. The evolutionary history also shows how the brain structures for consciousness came to be so diverse.

Figure 7.1 (plate 8) gives the time line for this chapter.[1] It shows that consciousness appeared in the Cambrian Period, about 540 to 520 million years ago (mya).[2] There is some uncertainty about the older limit, so a more conservative estimate is somewhere between 560 and 520 mya. Either way, it means that consciousness evolved from other nervous functions during the so-called Cambrian explosion, a special time in the history of life on Earth. Animal life never evolved faster than at this time.

Not long before the start of the Cambrian, the most advanced animals were simple ocean worms, whose nervous system comprised mostly a diffuse nerve net without any brain or complex

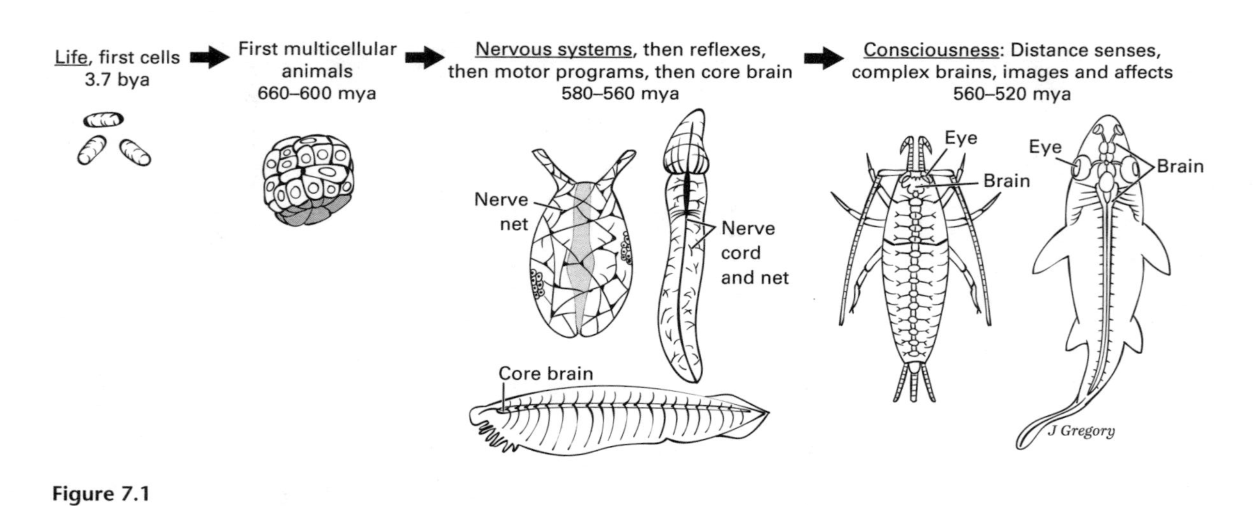

Figure 7.1
Stages in the evolution of consciousness, as covered in this chapter. Note the long interval between the dawn of life at 3.7 bya and the appearance of consciousness in the Cambrian explosion at 540 to 520 mya. The arthropod at right is a bristletail insect, and the fish is a shark. See plate 8.

sensory organs (fig. 7.2).[3] These worms fed on a rich layer of scum, called a microbial mat, that covered much of the shallow ocean floor across the globe. We know this from worm trails in the fossilized remains of the mats. The trails show that the worms' behavior was simple, just following the food sources (or desired mates) that they sensed and smelled.[4] The worms had attained the "reflex" stage of box 6.2. Other animals living at the time were even simpler, leaf-like forms, swaying in the ocean currents (fig. 7.2). Not much was going on, and few behavioral interactions occurred among animals.[5] Yet just a few million years later, the ancestral worms had evolved into all the thirty-plus phyla of animals that ever lived (fig. 7.3), interacting in heavy competition especially as predators versus prey, and most of these phyla have members that survive to this day.

Why the rapid change? The most popular theory says that evolution accelerated because the first predator animals evolved. That is, one species of worm gained the ability to eat worms of other species, perhaps after passing through an intermediate stage of scavenging disintegrating carcasses in the microbial mats.[6] In response to the new predators, through natural selection, the prey species evolved survival strategies that ranged from hiding tactics, to body armor, to keener senses and faster locomotion for detecting and evading the predators. In response, the predators evolved their own better senses, locomotion, and body weapons to detect, catch, kill, and eat prey. Thus these life forms entered into a kind of arms race, as only those predators and prey who escalated their offensive and defensive tactics survived. It is a fact of nature that larger animals attack and defend better, so many species increased in size. Numerous and varied survival strategies proved successful, leading to a rapid increase in animal diversity, behaviors, and body size.

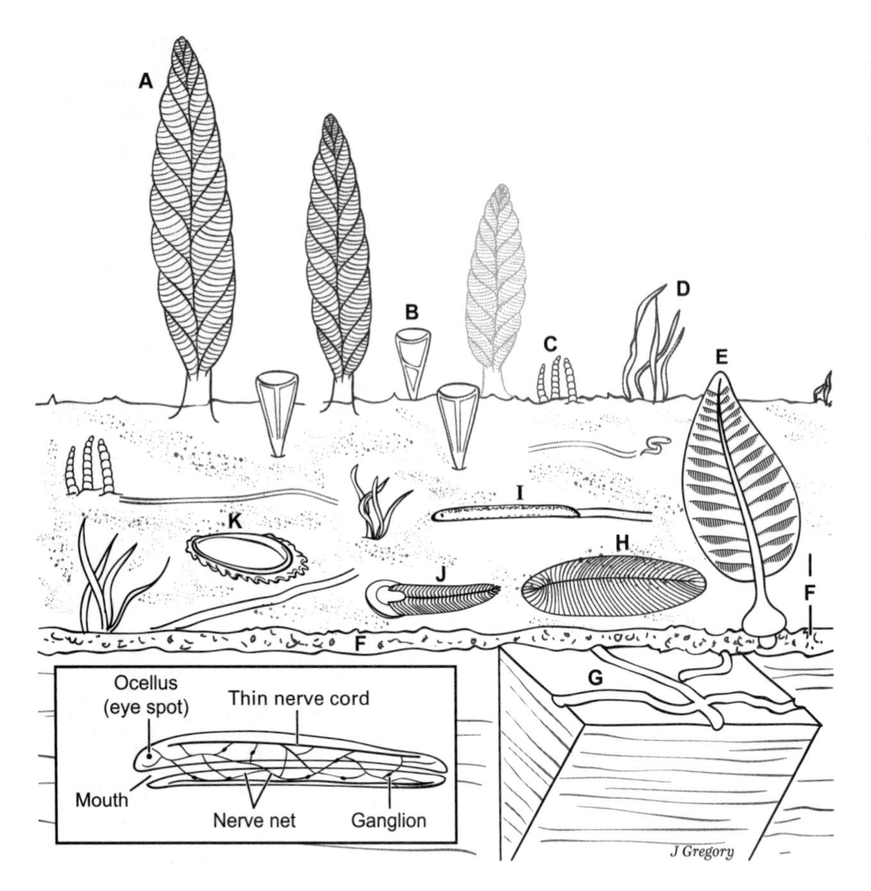

Figure 7.2
Life on the Precambrian seafloor shortly before the Cambrian, about 555 mya. Inset at lower left is a reconstruction of the ancestral worm that lived in that community, an enlargement of *I*. Key: (*A*) leaf-like animal, *Charniodiscus*; (*B*) sponge, *Thectardis*; (*C*) tubular body fossils, *Funisea dorthea*; (*D*) seaweed; (*E*) leaf-like "rangeomorph," *Charnia*; (*F*) the microbial mat; (*G*) worm burrows on underside of the microbial mat; (*H*) mobile segmented animal, *Dickinsonia*; (*I*) bilaterian worm foraging on microbial mat; (*J*) segmented animal, *Spriggina*; (*K*) possible mollusk, *Kimberella*.

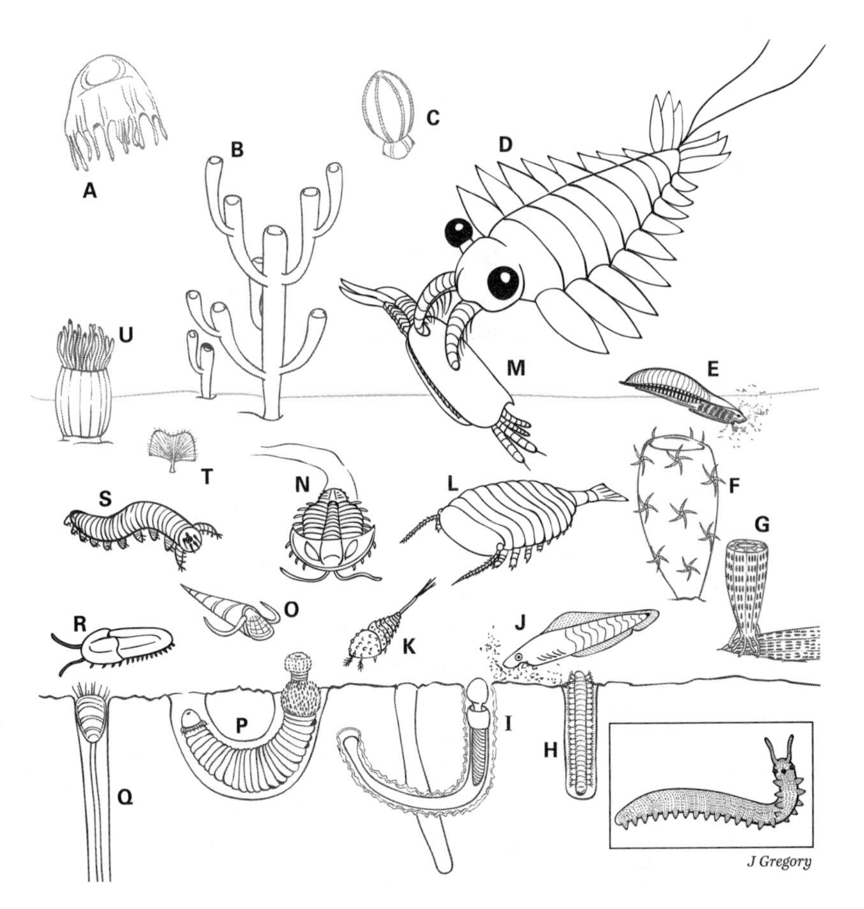

Figure 7.3

Cambrian seafloor, 520–505 mya, showing the explosion of animal diversity since the late Precambrian. Inset at lower right shows a modern velvet worm, related to the lobopodian worm at *S*. Key: (*A*) jellyfish of family Narcomedusidae; (*B*) sponge, *Vauxia*; (*C*) sea gooseberry, ctenophore *Maotianascus*; (*D*) anomalocarid arthropod-relative, *Amplectobelua*; (*E*) near-vertebrate *Haikouella*; (*F*) sponge, *Chancelloria*; (*G*) sponge, archaeocyathid; (*H*) annelid worm, polychaete *Maotianchaeta*; (*I*) hemichordate worm, *Spartobranchus*; (*J*) vertebrate fish, *Haikouichthys*; (*K*) arthropod, *Habelia*; (*L*) arthropod, *Sidneyia*; (*M*) arthropod, *Branchiocaris*; (*N*) arthropod trilobite, *Ogygopsis*; (*O*) brachiopod relative, hyolithid; (*P*) priapulid phallus worm, *Ottoia*; (*Q*) brachiopod, *Lingulella*; (*R*) arthropod trilobite, *Naraoia*; (*S*) lobopodian worm, *Aysheaia*; (*T*) brachiopod, *Diraphora*; (*U*) sea anemone, *Archisaccophyllia*.

This new environment, with so many animals and so much activity, sent out many more signals: sights, sound vibrations, smells, things to touch and taste. The two most active groups of animals, the incipient arthropods and vertebrates, took advantage of all this transmitted information by greatly elaborating their distance senses. That is, they evolved keen eyes, ears, noses for smelling, and better touch senses.[7] The early arthropods, with their jointed, manipulative limbs, included the main predators. The first vertebrates were harmless little fish that feasted on the remains of the microbial mats and swam so fast that they readily escaped predators (fig. 7.4; plate 9). Processing the tidal wave of new signals reaching their distance senses, the nervous systems and brains of both groups became more complex. The original reflexes, simple motor programs, and integrative core brains elaborated to become processors that merged all kinds

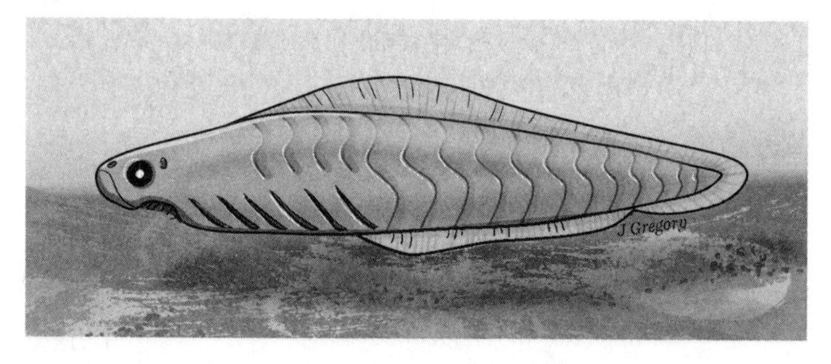

Figure 7.4
Haikouichthys, the oldest known fossil vertebrate from the Cambrian rocks of China, 520 mya. This was an inch-long jawless fish, with elaborate eyes, ears, a nose for smell, and a streamlined body for effective swimming. It could not have been a predator; any tentacles over the mouth were soft, so it had no teeth, claws, or grasping arms to obtain prey. See plate 9.

of sensory inputs into conscious mental images, with affects, for efficiently driving and directing motor actions in three-dimensional space. In short, the first vertebrates and arthropods evolved all the special features of consciousness so as to survive in the newly dangerous, highly complex environment of the Cambrian seas. Elaborate nervous systems use a lot of energy, and so do the many locomotory movements of the active lifestyle, requiring lots of food. Therefore most of the other new groups of animals evolved cheaper survival strategies and did not evolve consciousness. They burrowed, for example, or grew protective shells (fig. 7.3).

In the cephalopod ancestors of squid and octopuses, consciousness also evolved with an increase in predatory ability, but not at the same time as in the predatory arthropods. That is, fossils indicate that the cephalopods evolved later, at the end of the Cambrian (490 million years ago), after the bang of the Cambrian explosion (540–520 million years ago) was over. As mollusks, cephalopods must have come from a relatively inactive, heavily shelled ancestor, then secondarily switched to active hunting upon evolving their tentacle arms and jet propulsion. We cannot even be sure the earliest cephalopods were conscious, because all our evidence for their consciousness comes from the octopus-cuttlefish-squid group that appeared much later, about 275 million years ago, according to the fossil record.[8]

Thus, as illustrated in figure 7.5, we know that the vertebrates, arthropods, and cephalopods all evolved consciousness independently, the first two groups during the predatory arms race of the Cambrian explosion, and the cephalopods later.

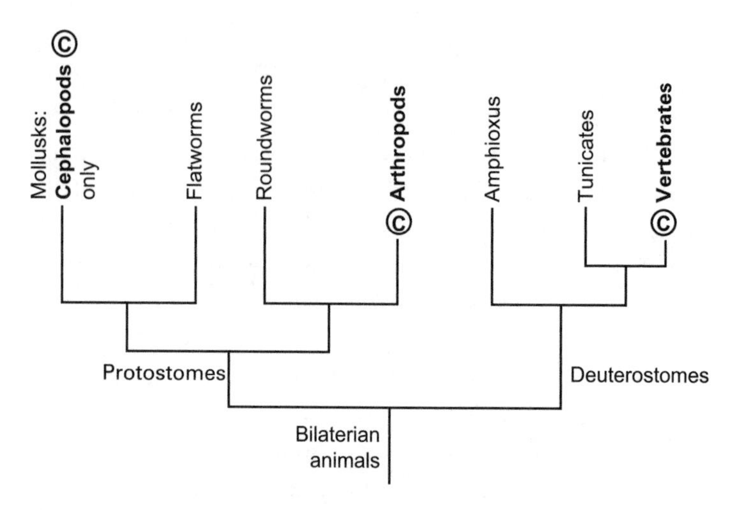

Figure 7.5

Consciousness evolved independently three different times. Phylogenetic tree showing the distant evolutionary relationships of the three groups of animals with consciousness (C). For simplicity, many of the other phyla of invertebrates have been left out. The phylum name for flatworms is Platyhelminthes, for roundworms is Nematoda, for amphioxus is Cephalochordata, and for tunicates is Urochordata.

The Historical Steps in the Evolution of Consciousness

So far we have considered the evolution of consciousness as revealed by the fossil evidence and by modern animals that descended from the Cambrian explosion. During that explosion, the neural transition accelerated but was continuous as the general features of life and simpler neural traits elaborated into the special features of consciousness.

Now, to provide a fully naturalized explanation for consciousness, our evolutionary scenario will show that the first appearance of the conscious animals occurred simultaneously with the evolution of their capacity to create sensory mental images and affects.

The Evolutionary Transition to Sensory Mental Images

The most basic forms of sensory responsiveness to the environment were reflexive and innate, performed with nothing more than simple motor programs. These primitive responses characterized the early, nonconscious bilaterian animals such as ancestral worms. We saw these basic forms in figure 6.2, which shows the body withdrawal of a roundworm, and in figure 6.3, which shows the repeated feeding movements of a snail-like sea hare, two animals we judged to lack consciousness. Notice from the circuit diagrams in these figures that we can describe these reflex-level responses in an entirely objective way and from an entirely objective viewpoint without any gaps in their explanation.

In the Cambrian, however, when increasing brain complexity and the evolution of special features first turned reflex-level processing into mental images and affects, the aspects of subjectivity were created. This is because the images and affects are private, referred, unified, mentally causal, possess qualia, and entail all four explanatory gaps that characterize subjectivity (box 2.1).

As noted in chapter 6, the special feature of *complex neural hierarchies* builds mapped representations of different objects in the environment from multiple elaborate senses and joins all these inputs into a unified conscious image. In the first vertebrates, these sensory images were enabled by a great expansion of the distance senses.[9] As mentioned, we hold that vision was the first sense to elaborate in the vertebrates, with the appearance of the camera eyes (chap. 3). We follow this vision-first idea not only because the oldest known fossil fish has a camera eye (plate 9) but also because the sharply focused images from such eyes can yield more information about the outer world than can any other sense.[10] That is, only vision tells the positions, borders, and movements of many distinct objects, from the farthest distances, as well as telling which object lies in front of another in space. Alternate claims that smell was the first sense to elaborate probably do not apply to the vertebrates because their closest invertebrate relatives (amphioxus and tunicates) have no obvious smell organs but do have eye rudiments (chap. 3).[11] In the vertebrate line, the first mental images were visual, but input from the other new senses was soon added to make a multisensory map (in the brain's optic tectum).

What about the evolution of mental images in the arthropod and cephalopod lines? The nearest living relatives of arthropods are the forest-dwelling onychophorans (velvet worms) of the southern continents, who had many relatives on the Cambrian seafloor called *lobopodian* worms (fig. 7.3).[12] The brain of the velvet worms is comparable to that of arthropods, except it lacks the big visual-processing centers, as these worms have small eye cups instead of the large, compound, image-forming eyes of arthropods. Clearly, vision and the capacity for visual images exploded at the dawn of the arthropods, just as it did at the dawn of the vertebrates.[13]

In cephalopod squid, octopuses, and cuttlefish, the brain evolved from a small pair of cerebral ganglia such as occur in the heads of some other mollusks, specifically in gastropods (snails and their kin).[14] The gastropods' brain ganglia are so much simpler, however, that it is hard to reconstruct the evolutionary steps that led to the cephalopod brain. Nonetheless the eyes and visual centers of the cephalopod brain are so much larger and more complex than in any other mollusk (fig. 5.1) that visual expansion must have been important to the evolution of mental images in the cephalopods.

From these findings, we see that the appearance of image-based consciousness relied heavily on the evolution of sharp vision in vertebrates, arthropods, and cephalopods.

The Evolutionary Transition to Affective Awareness

The neural evolution of affective consciousness, of positive and negative affects, is easier to trace in the vertebrates than in the less-studied arthropods and cephalopods. Recall the affective brain systems in vertebrates from figures 4.1 and 2.3, where they are shown to include the amygdala, basal forebrain, periaqueductal gray, and more.

In the beginning, the ancestral worm had only a simple *integrator circuit for incentives,* such as exists in the small brain of today's sea slugs. As shown in figure 4.3 (plate 5), this kind of circuit assigns positive or negative valence (value) to the incoming sensory information and then tells the nearby *central pattern generators* to start or stop a motor response, as is appropriate for the assigned valence.

Then, when the first vertebrates evolved global operant learning—the ability to learn brand-new behaviors from experience based on rewards and punishments (chap. 4)—the positive and negative valences first became conscious and then became more varied. For example, negative affect came to include intense fear as well as the original dislike, and positive affect split into feelings of food joy and sexual lust. The ancestral valence-assigning circuit elaborated and split into multiple circuits or centers that were distributed across the brain (figs. 4.1, 4.2), with each center specializing in a different affective role. These separate affective centers remained strongly connected to each other, however, and to the premotor brain areas that chose motor actions (basal ganglia, hypothalamus), through which the affective centers continued to influence motor behaviors. The same developments seem to have occurred in the brains of arthropods, where separate affective centers also communicate with premotor regions.[15] The story was one of increasing complexity, but it seems to have peaked early in the history of both the vertebrate and the arthropod lines. We say early because lampreys have most of the affective brain structures present in all the other vertebrates, including humans (chap. 4), and because the fossilized brains of the earliest arthropods have the same structures and complexity as do the brains of living arthropods (chap. 5).

Historical Summary of the Stages Leading to Consciousness

In documenting the smooth evolution of consciousness, it helps to summarize the important steps, along with the adaptive benefits and advances at each step. To this end, we now consider plate 8 (fig. 7.1) step by step. At 3.7 billion years ago (bya), the first bacterium-like cell had a private body with a border to the outside world and was the first life. Much later, around 660 to 600 million years ago (mya), cell colonies led to multicellular animals. Now the colony, not the cell, was the embodied organism. This increased the body size of the entire organism, and its adaptive benefit was that larger organisms avoid predation. Next the many cells in the body differentiated into distinct tissues and organs for an efficient subdivision of labor. One of the tissues was neural, meaning a nervous system had appeared. Its neurons allowed distant regions in the body to communicate with one another, by means of reflexes. This neural communication allowed quick reactions of the whole body to stimuli and also improved coordination of all the internal body functions. This reflex stage led to the more complex reflexes and simple motor programs in ancestral worms, all of which is attested by fossil worm trails at 580 mya. Next came a core brain for basic survival functions, arousal, and better inner-body homeostasis in both the prearthropods and the prevertebrates. Finally, consciousness evolved, with the distance senses and other special features, in the first arthropods and first vertebrates, around 560 to 520 mya.[16] As consciousness evolved, it offered special adaptive advantages.

The Adaptive Value of Consciousness

We have already introduced some of the benefits of having consciousness, mainly its allowing a body to operate in complex, three-dimensional space. Now we expand on the survival value of experiencing both mental images and affects (box 7.1).[17] First, consciousness, with its unified central stage of images and affects, seems to be the best way to organize and integrate vast amounts of incoming sensory information for action choice. It is more efficient and requires far fewer neurons than would be prespecifying the responsive action for each and every type of stimulus likely to be encountered: no brain could hold that many prespecified reflex arcs or precise motor programs, and each of them would be too rigid and inflexible.[18] We can explain this greater efficiency of consciousness from the perspective of evolutionary history. In the first vertebrates and arthropods, when so many new kinds of sensory inputs began feeding into the brain, they came to be organized into multisensory mapped images, in which any ambiguities or apparent inconsistencies between the signals from the different senses could be resolved.[19]

Box 7.1

Adaptive advantages of consciousness

- It efficiently organizes much sensory input into a set of diverse qualia for action choice. As it organizes them, it resolves conflicts among the diverse inputs.
- Its unified simulation of the complex environment directs behavior in three-dimensional space.
- Its importance ranking of sensed stimuli, by assigned affects, makes decisions easier.
- It allows flexible behavior.
- It allows much and flexible learning.
- It predicts the near future, allowing error correction.
- It deals well with new situations.

Furthermore, the many different sensed items in the image were ranked according to their importance, by giving the most important items the most attention and assigning to them the strongest affects (emotions). The important items in the images, in turn, determined which motor program was ordered and carried out (e.g., a rich food source induced the feeding program, a predator induced the fleeing program, etc.).[20] Importance ranking also lets the conscious animal attend to *specific targets*, and the assigned affects tell the animal which targeted items to approach and which ones to avoid.[21]

By contrast, nonconscious animals that rely only on reflexes and motor programs cannot effectively process as much sensory information. This limits their range of activities, and because they have no map of where they are, they cannot perform guided or proactive behaviors.[22] Conscious animals can move almost anywhere and manipulate many objects, even in a jam-packed three-dimensional habitat with obstacles.

Consciousness allows more behavioral *flexibility* than do nonconscious systems. The many different items in the mapped conscious image are readily reprioritized by redirecting attention to, and reassigning affects from, one item in the perceived world to another, thus easily changing the resulting responses. This is behavioral flexibility.[23] Consciousness can also allow behavioral flexibility even when it is not mapping a lot of complex stimuli: that is, the more purely *affective* type of consciousness, which takes unmapped or relatively simple stimuli and ranks them by valence, can be easily and flexibly redirected so that, for example, a sudden threatening noise changes a curiosity-driven approach behavior (+) to a fearful, directed, prolonged flight (–). So does a sudden smell of decay and death. Other benefits of consciousness are that it allows a lot of learning (UAL; see chap. 4) and short-term predictions (fig. 6.5), both of which provide enormous adaptive value.

In this chapter, we have found that consciousness is a natural phenomenon with a smooth but "punctuated" evolutionary history (fig. 7.1; plate 8) from the basic features of early, one-celled life to reflexes and then to complex nervous systems with special neural features that formed mental images and affects. We have also considered the adaptive benefits of evolving both the aspects of consciousness, its images and affects. We traced the evolution of these aspects most thoroughly for the vertebrates. In the arthropods and cephalopods, we could reconstruct the evolution of images well enough to see that it probably followed the same path as in vertebrates, with the early evolution of vision and visual images being especially important.

Having laid out all the steps in the evolution of consciousness in a natural context without any gaps, in the final chapter we try to integrate our analysis of the neurobiology and evolution of consciousness with the explanatory gaps and subjectivity.

8 Naturalizing Subjectivity

Thus far we have spent most of the book identifying the explanatory gaps and the common biological and neurobiological features needed for the creation of the various forms of sensory consciousness (chaps. 2–6), and showing how these features naturally evolved (chap. 7). In this final chapter, we integrate this information and try to explain one of the great mysteries of science: the relationship between the brain and subjectivity.

Recall that in chapter 1 we said that our naturalized theory of consciousness and our explanation of the mysterious explanatory gaps stem from three principles: first, consciousness and subjectivity are built on the unique features of life; second, while consciousness is built on these general features of life, it also depends on additional simple and complex reflexes and core brain functions, to which are added *special neurobiological features* that are in fact *unique* to conscious brains; and third, the unique nature of consciousness creates the divergence between the subjective and objective points of view.

Our model also proposes that the gradual accretion of novel properties over millions of years during the evolutionary progression from nonlife to life to reflexes to consciousness created both consciousness and the explanatory gaps. We now present our complete model from the beginning of life to the creation of subjectivity (fig. 8.1; plate 10).

Life Is Fundamental to Explaining Consciousness, Subjectivity, and the Explanatory Gaps

At the first rung of our model is the principle that consciousness and subjectivity are fundamentally grounded in general life functions. By this we mean that consciousness is inextricably and irreducibly a part of the *physical* life of the organism.[1]

We find support for this proposal in the striking commonalities between many of the features of the life of the organism and sensory consciousness (chap. 6). Namely, both life and consciousness are embodied, both are processes, both are unique system features of the animal, and both life in general and sensory consciousness in particular are the natural result of a complex system's subparts and their systemic—especially hierarchical—interactions.

One point is particularly important for explaining the subjectivity and explanatory gaps at the top of the hierarchy: because both life and consciousness are system features of *embodied* organisms, it follows that all the features of both life and consciousness are personal and unique to the living individual, *so the subjectivity of consciousness is already inextricably rooted in the personal life of the animal*. Therefore any explanation of the explanatory gaps must take into account that, even at the level of the general features, the subjectivity of the later-evolving consciousness was already inherent.

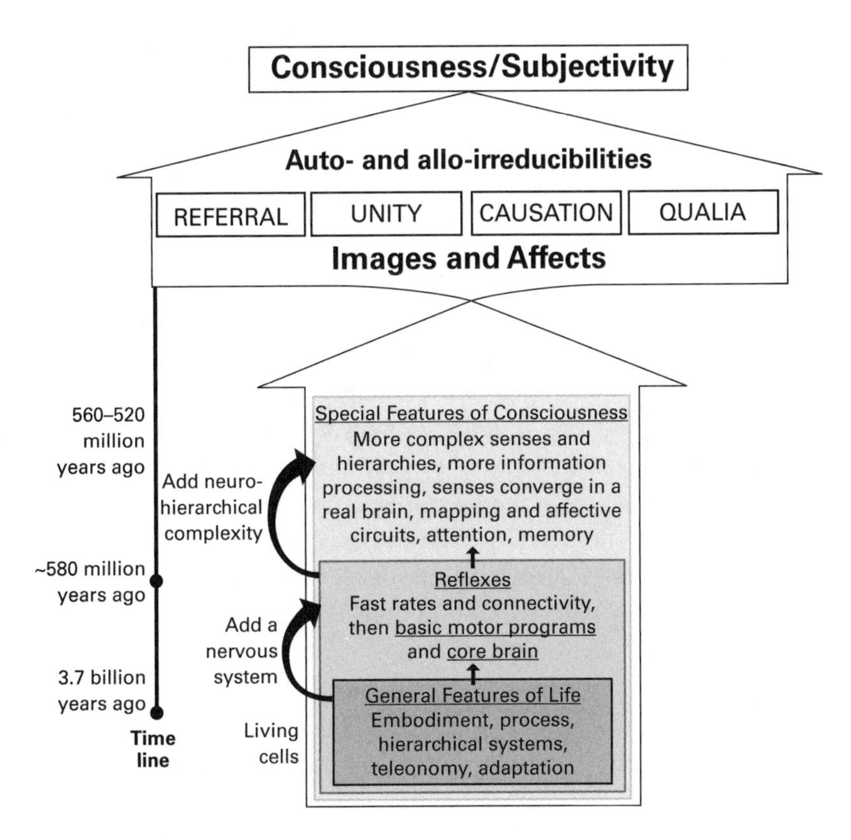

Figure 8.1

Summary of our naturalistic solution to the questions raised by consciousness and subjectivity, from the evolutionary steps, to the necessary biological and neural features, to the components of subjectivity (*top*) that we claim to explain. The four neurologically subjective features of consciousness (referral, etc.) appear in the rectangles above. We also call these the four explanatory gaps and gap features. See plate 10.

Reflexes and Core Brain Functions Bridge the General and Special Features

Although we do not consider reflexes or core brain functions as conscious, their evolution provided the critical link between life, consciousness, and subjectivity (chap. 6). Among the critical new factors at this intermediate level are specialized neurons, neural coordination of an entire multicellular body, enhanced speed of neural processing, neural circuits, and the start of neurohierarchical complexity (see the amphioxus in figs. 3.5–3.6). We will consider later how the building of consciousness on the reflexes also predetermines and sets the stage for the explanatory gaps of referral, unity, and mental causation.

The Special Neurobiological Features Are Fundamental to Explaining the Unique Aspects of Consciousness and the Explanatory Gaps

Consciousness requires the special neurobiological features that are indeed unique to conscious brains. These new features, in combination with the life processes, reflexes, and core brain functions on which they are built, naturally create the sensory images and affects of subjective experience.

At this level of the special features (box 6.3, fig. 8.1, plate 10), we find the explosion of special senses and neuron types, the multitude of new neural processing subsystems, more neurohierarchical levels, more integration of information from the different senses, more reciprocal communications between the lower and higher levels, more effective attention, and more memory. From these features arise neurobiologically unique system properties of complex brains in a way comparable to how life is naturally created from the interactions of its subcellular and cellular components. Note that this explanation requires no new "fundamental properties" or unexplainable or novel principles of physics or nature, just the normal but unique mechanisms of life and neural systems.[2]

However, we also found that the neurobiological basis of primary consciousness, along with its accompanying subjective features like qualia, is remarkably varied both within individual brains and across the brains of different species from vertebrates to arthropods to cephalopods. To reinforce how diverse the neural contributors to consciousness are, let us reconsider a part of the vertebrate brain that we have only touched on: the wide-ranging reticular activating system and thalamus that contribute to attention, alertness, arousal, and wakefulness, without which consciousness and therefore subjectivity would not be possible (chaps. 3 and 6; fig. 4.1).[3] This immensely broad distribution of structures that contribute to consciousness—perhaps the broadest of all neural functions—ensures that the neural basis of consciousness will be at best poorly localizable within the brain.

Explaining the Gaps and Qualia

As we have noted, the explanatory gaps can be directly related to the general and special features. Now we will explain and fill these four gaps of subjectivity—referral, unity, mental causation, and qualia—by beginning our analysis with the general features of life.

The *referral* of exteroceptive mapped mental images ultimately comes from the relation of the organism to its environment. Start with the general feature of embodiment and its attendant system features that initially established a *nonconscious* organism with an internal (organism) versus external (environment) relationship ("Living cell" at bottom of plate 10). Higher up, at the level of the reflexes, external stimuli are automatically redirected out to the environment, further reinforcing the relationship of organism to environment. We see this phenomenon, for example, in the way any animal's body automatically withdraws from or pushes back against a harmful stimulus to protect itself. The integration of the organism versus the world is the very foundation of the further development of subjectivity. Next, in the progression from reflexes to consciousness, isomorphic maps are added to the foundation to create the referred sensory images that distinguish world from body in exquisite detail. This takes us to "referral" at the top of plate 10.

Conscious *unity* at the higher level of the special features is directly linked to reciprocal neural interactions that bind coded sets of sensory information together into a unified image or affect (top of plate 10). However, down at the more basic levels, all physiological life processes are integrated and unified to achieve homeostasis, and the reflexes are genetically prewired to create linked programs that effect unified actions. In short, the unity and integration that result from the special features (box 6.3) stem from the unified systemic features of life and the reflexes at the lower levels. Again, we have derived a gap feature at the top of plate 10 ("unity") from the physical features lower down in the figure.

Mental causation is the easiest of the gap features to relate all the way back to the first embodied life (plate 10). Any living cell reacts to its surrounding environment—by moving, taking up food, or depositing its wastes, meaning that its reaction *causes* effects on that environment. So do an individual's unconscious

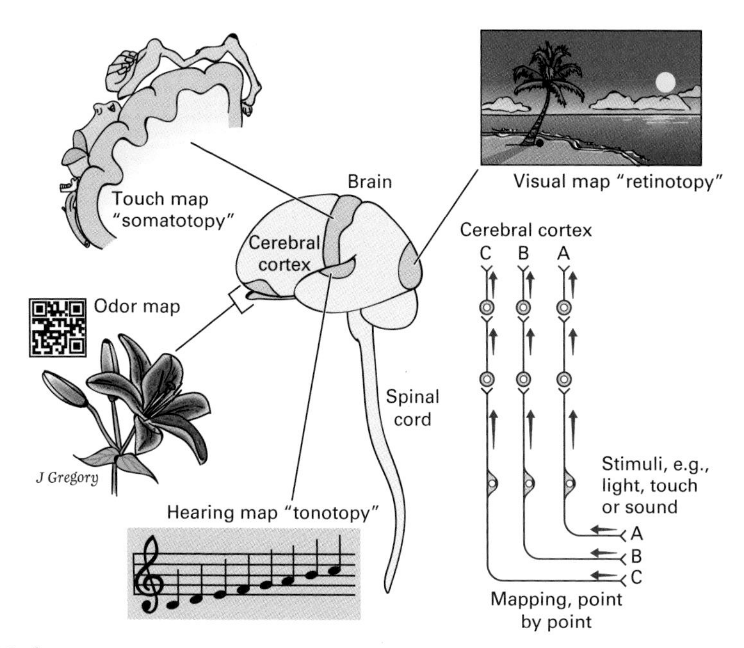

Plate 1

Mapped isomorphic organization of the exteroceptive sensory pathways. Each sensory pathway of several neurons (*right*) is a hierarchy that carries signals up to the brain, keeping a point-by-point mapping (A, B, or C) of a body surface, a body structure, or the outside world. This mapping leads to the mapped mental images that are drawn around the brain. The touch map of the body (*upper left*) includes a cut section through the folded cerebral cortex. The bar code associated with the flower at left shows that each complex odor has its own, coded scent signature.

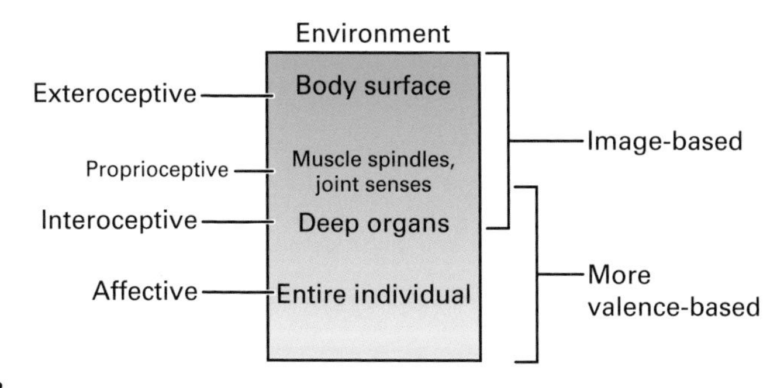

Plate 2

The subtypes of consciousness. *Left side*: Three domains of sensory consciousness shown as a continuum: exteroceptive, interoceptive, and affective. Proprioceptive consciousness is also shown, with less emphasis. *Right side*: In some parts of this book, we simplify the domains down to just two: image-based and affective (valence-based). Image-based consciousness includes exteroceptive, proprioceptive, and the somatotopically mapped aspects of interoceptive. Muscle spindles and joint senses measure proprioception, the stretch produced in our body's muscles and joint capsules as we move.

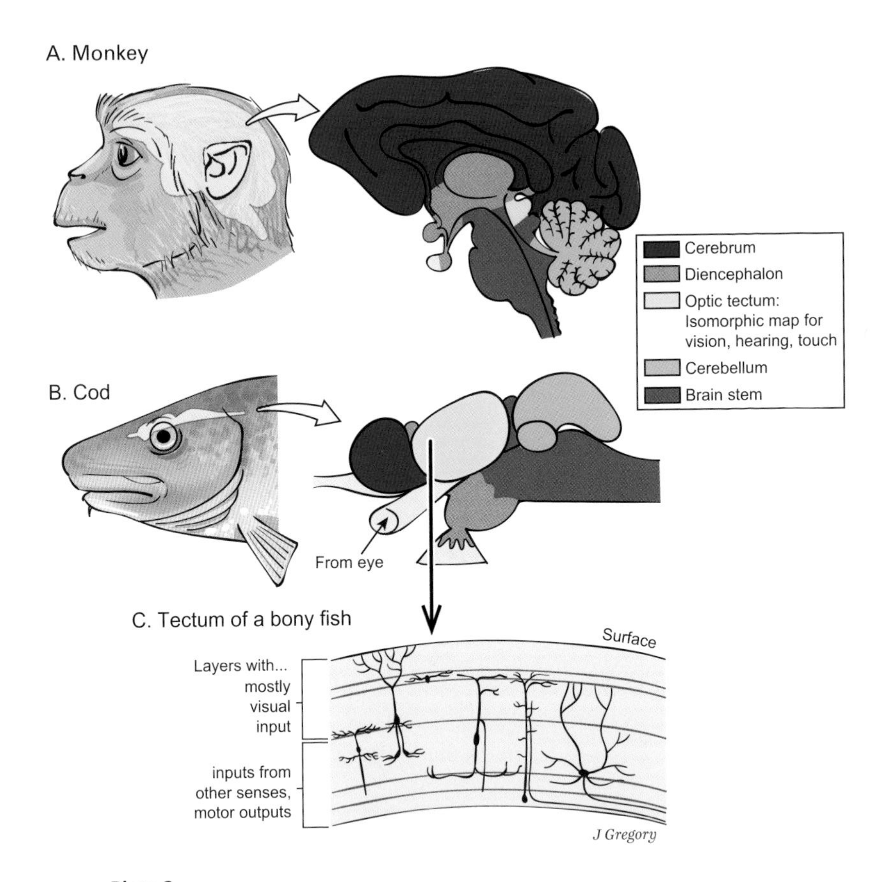

Plate 3

Optic tectum of the brain of a mammal (*A*) and a bony fish (*B*), with a section through the fish tectum (*C*). *C* shows some of the kinds of neurons in the tectum. We make the case that the fish tectum forms conscious images, but the mammal tectum does not. Brain in *A* is cut in half lengthwise; brain in *B* is uncut and whole.

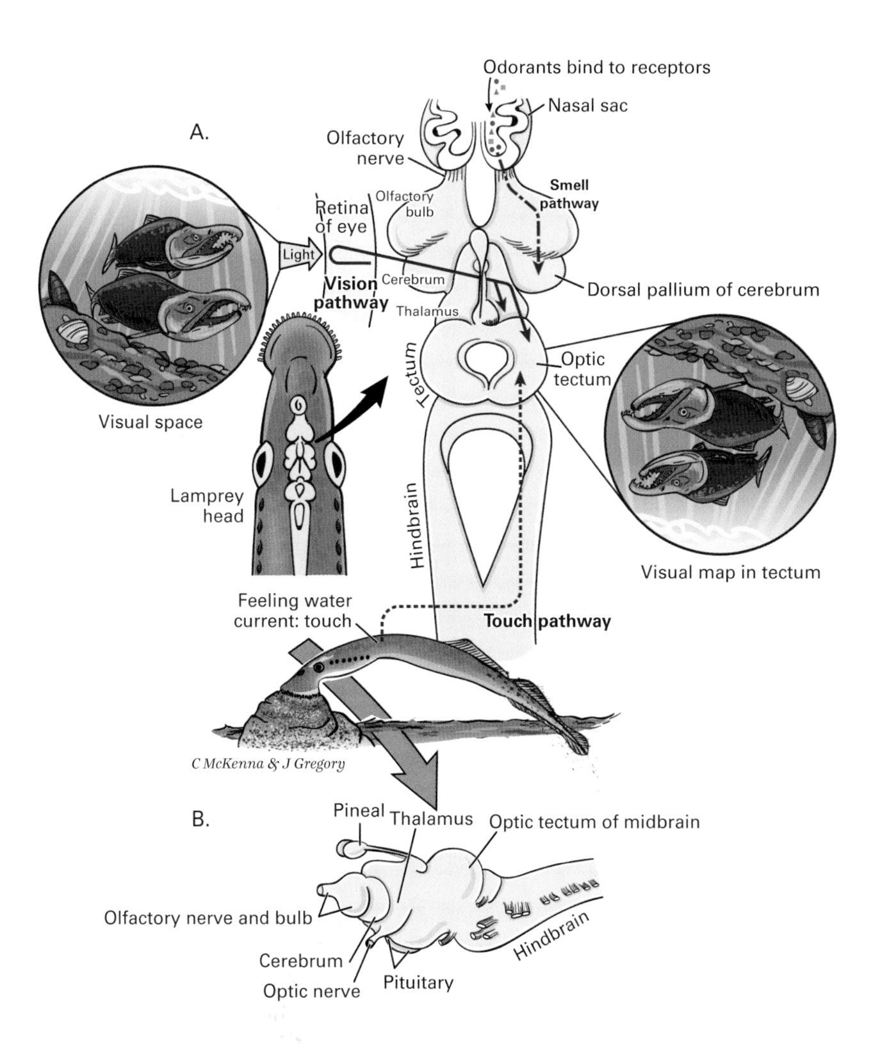

Plate 4

Sensory pathways and the brain of the lamprey, a jawless fish. A lamprey is shown sucking on a rock in a streambed, near the middle of the figure. Brains are viewed from above (*A*) and from the side (*B*). As shown in *A*, most sensory pathways reach the optic tectum, where they form a mapped (isomorphic) representation. This image shows two salmon in a spawning run in the same stream.

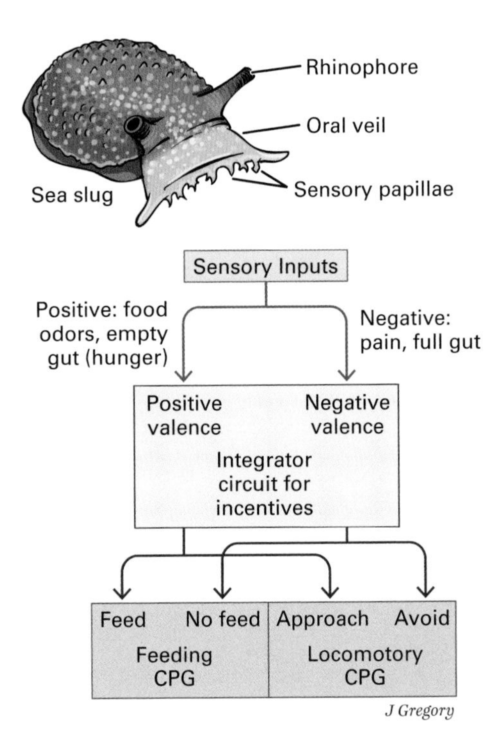

Plate 5

Basic valence circuit in the sea slug (*Pleurobranchaea californica*), which may show the simplest ancestral type. *Top*: the slug senses food and other outside stimuli with its sensitive oral veil and also senses the emptiness or fullness of its gut. *Middle*: these stimuli are assigned positive or negative valence in the integrator circuit for incentives, which tells the animal's central pattern generators (CPG) to start or stop the feeding and locomotory movements (*bottom*).

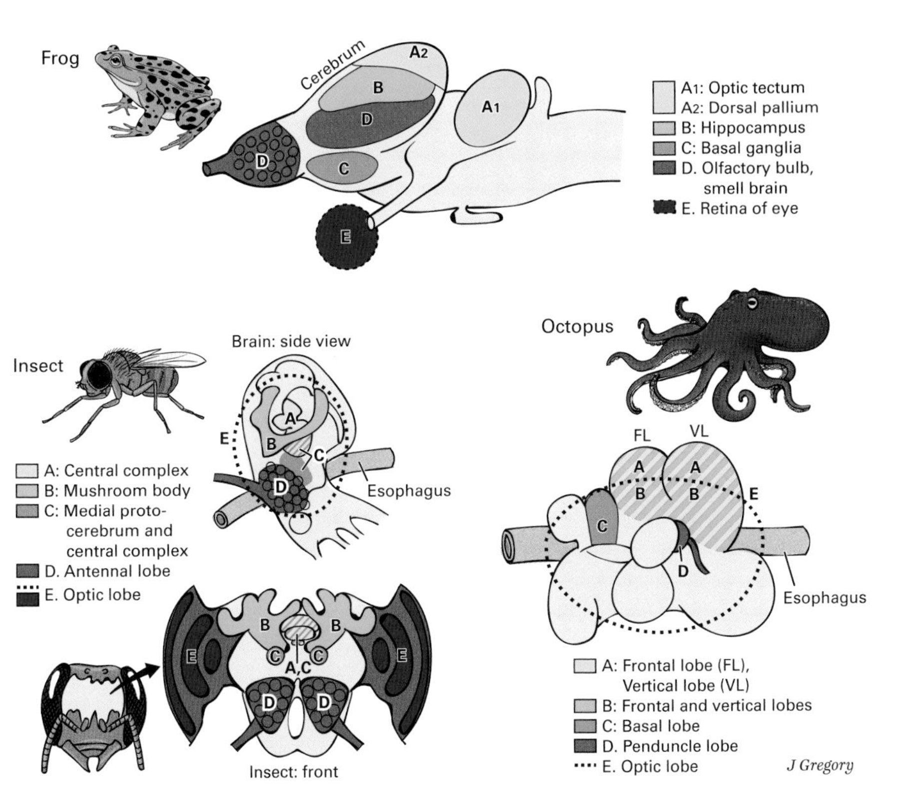

Plate 6

Brains of an insect and an octopus compared to brain of a vertebrate (frog). Mostly shown in side view. Regions with similar functions for consciousness are colored similarly in the three kinds of brains. For more information, see Barron and Klein 2016; Shigeno 2017. Despite the functional similarities, the three brains evolved independently of one another from the simple nervous system of a brainless ancestral worm (chap. 7). The general code: A, image-based consciousness; B, memory; C, premotor center; D, smell processing; E (with the dotted outlines), visual processing.

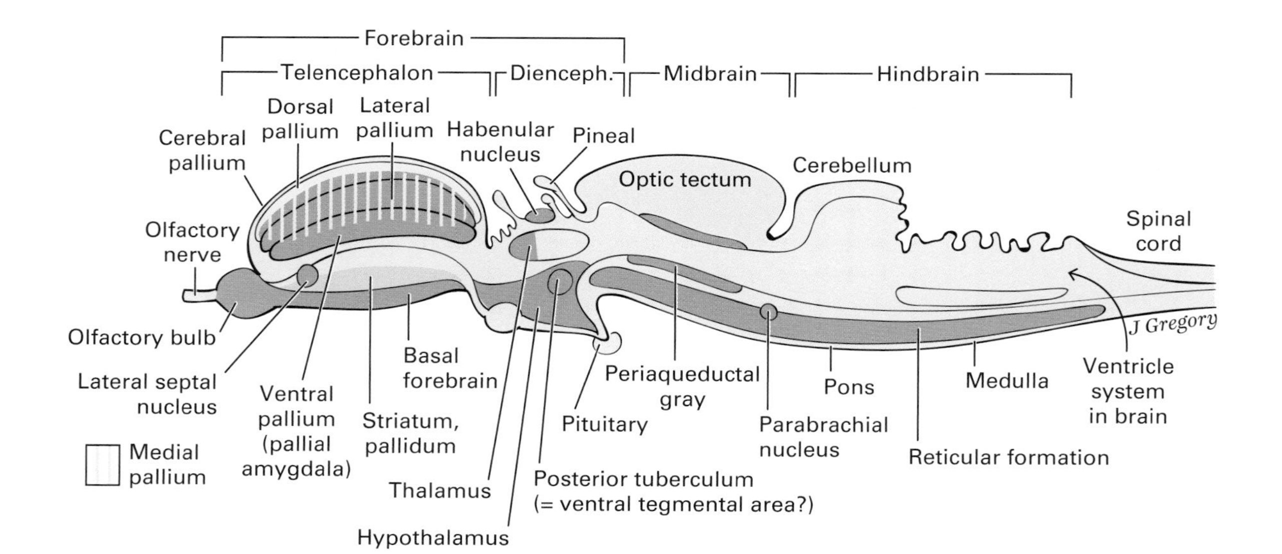

Plate 7

Regions of the vertebrate brain for image-based consciousness versus affective consciousness. These regions are mostly different, despite some overlap and much intercommunication. *Blue*, regions for affects; *yellow*, regions for mapped images.

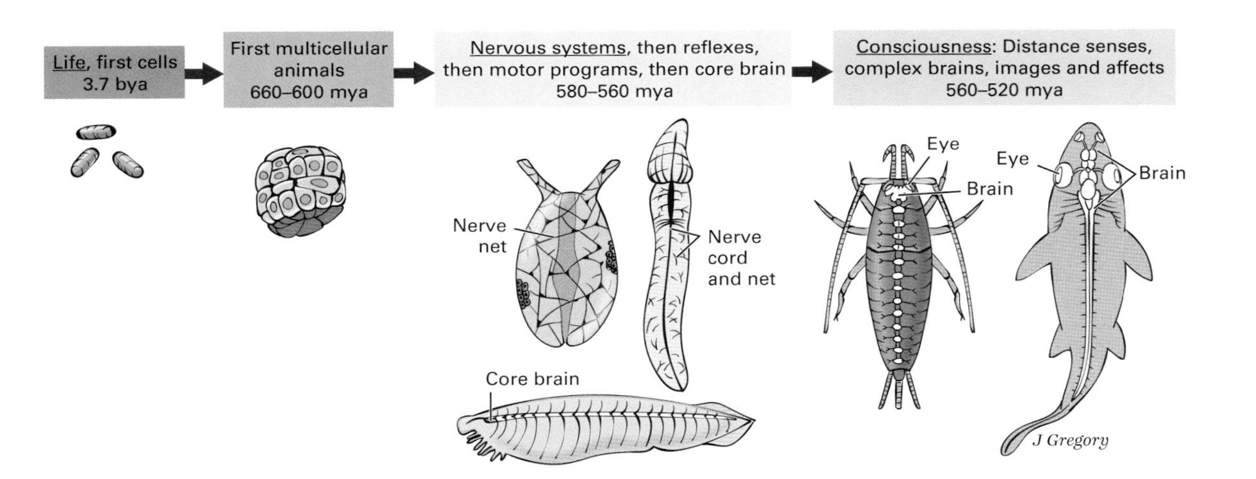

Plate 8

Stages in the evolution of consciousness, as covered in this chapter. Note the long interval between the dawn of life at 3.7 bya and the appearance of consciousness in the Cambrian explosion at 540 to 520 mya. The arthropod at right is a bristletail insect, and the fish is a shark.

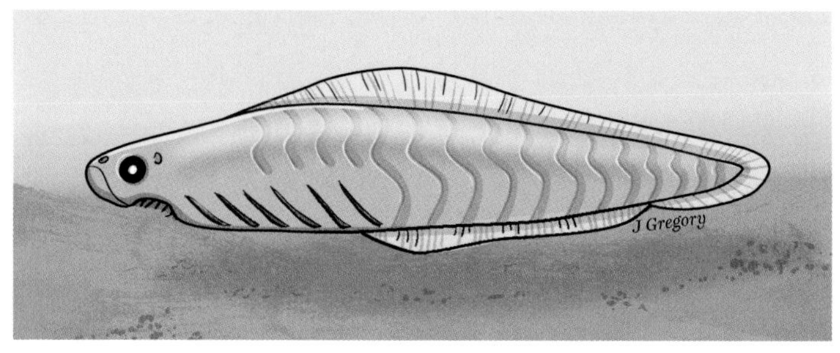

Plate 9

Haikouichthys, the oldest known fossil vertebrate from the Cambrian rocks of China, 520 mya. This was an inch-long jawless fish, with elaborate eyes, ears, a nose for smell, and a streamlined body for effective swimming. It could not have been a predator; any tentacles over the mouth were soft, so it had no teeth, claws, or grasping arms to obtain prey.

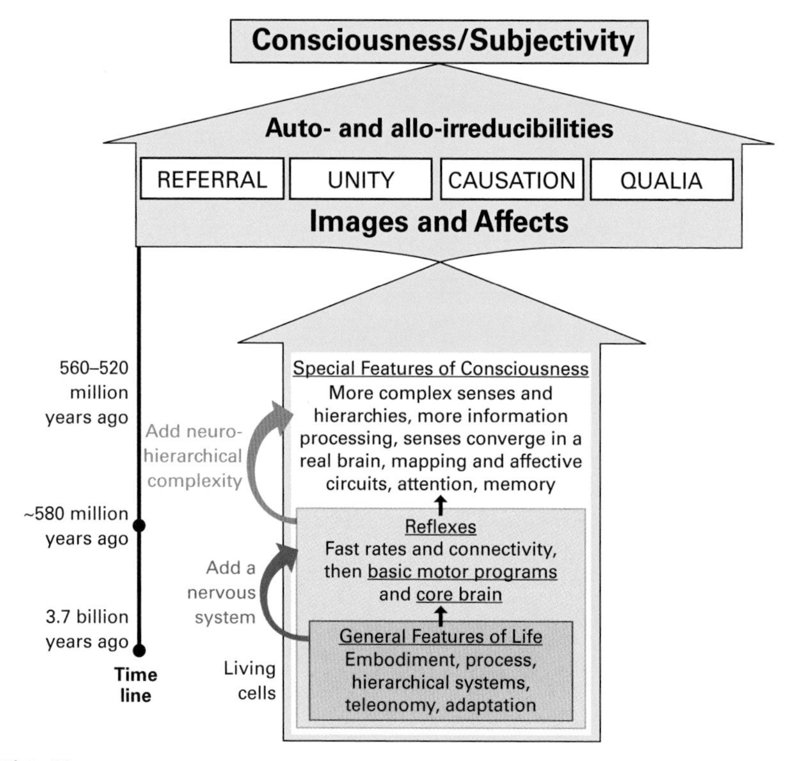

Plate 10

Summary of our naturalistic solution to the questions raised by consciousness and subjectivity, from the evolutionary steps, to the necessary biological and neural features, to the components of subjectivity (*top*) that we claim to explain. The four neurologically subjective features of consciousness (referral, etc.) appear in the rectangles above. We also call these the four explanatory gaps and gap features.

reflexive actions and motions. And conscious mental causation certainly fits this mold of an individual body acting on the environment: it is obvious that I cannot move your arm with my thoughts; I can only move my own arm.

But how do we explain *qualia,* the most perplexing of the explanatory gaps? As defined in chapters 1 and 2 qualia are the subjectively "felt" qualities of sensory consciousness. We think it is fair to say that many—even most—investigators consider qualia to be the most difficult mystery of consciousness. As Francis Crick and Christof Koch state:

> The most difficult aspect of consciousness is the so-called "hard problem" of qualia—the redness of red, the painfulness of pain, and so on. No one has produced any plausible explanation as to how the experience of the redness of red could arise from the actions of the brain. It appears fruitless to approach this problem head-on.[4]

We feel that the problem of explaining qualia, while complicated, is not insurmountable. First, as presented above in the "Life Is Fundamental" section, we find that all the features of consciousness, including qualia, come ultimately the life of the animal. To make this point for qualia, consider the following comparison. Respiration is a broadly distributed physiological function whose large-scale (macroscopic) processes mainly involve the lungs and heart. That is, they involve inhaling air, and the heartbeat that propels blood to the lungs. These macroscopic functions provide the necessary oxygen for the body's cells and for the subcellular processes in the mitochondria where the biochemistry of respiration ultimately takes place. Thus while the macroscopic and biochemical processes of respiration are quite dissimilar, they have in common that both are integrated parts of the living respiratory system.

Now consider qualia. The vertebrate brain on the macroscopic level functions very differently from the neurons and the synapses on the cellular level. But what they all have in common, what is essential to the system that produces consciousness and subjectivity, is that they are all constituents of the integrated system of the life of the animal. In this sense, qualia are part of the life functions of the whole nervous system and brain. Therefore the experience of the redness of red is a living process, as is the painfulness of pain, and so on.

As modeled in plate 10, qualia are now seen as integrated living processes of certain complex brains. Whether we are talking about such neurobiologically diverse feelings as red, pain, hunger, or happiness, *qualia cannot be dissociated from life processes in general*. Qualia in this sense are *alive* in the same way that a cell is alive, or a heart is alive, or a person is alive. When viewed in this light, the widely varying special features that are essential to the creation of images and affects—whether they require isomorphic mapping, hierarchically organized neural interactions, valence neurons, or other special features—are still living system features and processes.[5] So the subjective gaps that appear at the top of our model in fact have their origins at the bottom.

Furthermore, the creation of qualia also involves two other gaps that we have already explained, namely, referral and unity. Exteroceptive and interoceptive qualia are *referred* out to the world and body, where the stimuli arise. And qualia must also be *united* to some extent. Sounds, colors, tastes, and affects are united by millions of integrated neurons and numerous hierarchical levels (special features) into each qualia-rich experience.

Observation, Subjectivity, and the Explanatory Gaps: Auto- and Allo-ontological Irreducibility

A third principle is critical to understanding the subjectivity of qualia, as qualia appear at the top of plate 10. This principle pertains to the disparity between first-person and third-person perspectives of consciousness. It actually consists of two reciprocally related principles that make subjectivity (experience) different from objective description (fig. 8.2).

The first determining principle is *auto-ontological irreducibility*,[6] which is a fairly technical term for a fascinating but straightforward concept. Auto-ontological irreducibility simply means that one's *subjective* consciousness never refers to the *objective* neurons that create it, meaning that consciousness cannot experience how its mental events are physically produced. As the philosopher Gordon G. Globus described this feature: "It does

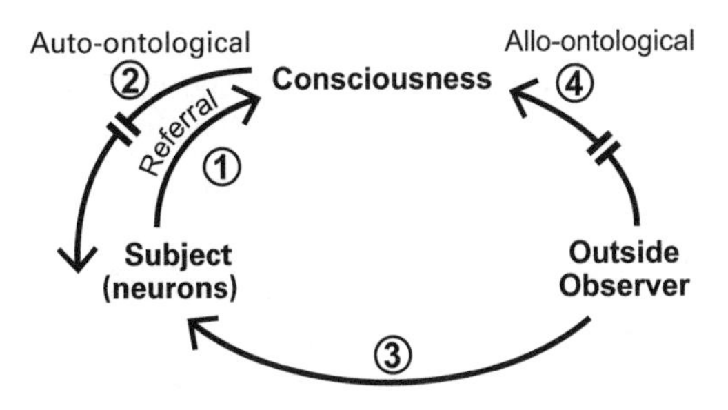

Figure 8.2
The auto- and allo-ontological irreducibilities of consciousness. The subject has access to its own referred conscious experiences (1), but no access to its own neurons that create the experience (2). This is auto-ontological irreducibility. An outside observer can objectively access the workings of the subject's neurons (3), but not the subject's conscious experiences (4). This is allo-ontological irreducibility.

not appear that the brain in any way codes or represents in any way its own structure. (The nervous system has no sensory apparatus directed to its own structure)."[7]

Additionally, even if it were possible for a subject to observe her or his own neurons objectively in the act of creating subjective experiences—for example, by using a hypothetical instrument called an autocerebroscope that would allow a person to look objectively at his or her own neurons[8]—these observations would simply amount to the same thing as an outsider's third-person observations, still encountering the barrier between one's own objective brain and one's felt, subjective experiences.

Auto-ontological irreducibility is explained in part by the way consciousness evolved, because even if it were possible, there is no adaptive advantage for one's "conscious" circuits to attend to themselves or to any other part of one's brain. More basic physiological processes do that instead, especially "babysitter" cells called *glia*, which monitor, nourish, and support the neurons and keep them functioning properly.[9] An animal's survival depends on its attention being directed to the outer world and to its body, so it would be a wasteful duplication of effort if sensory neural networks evolved to attend consciously to their own neurons.

The objective–subjective barrier also involves the unobservability of one's subjective states by another person. We call this *allo*-ontological irreducibility (*allo-* means "other"), and it is the flip side of auto-irreducibility (fig. 8.2).[10] Whereas auto-irreducibility means a subject cannot experience her or his own objective brain, allo-irreducibility means an objective outsider cannot directly observe or measure a subject's experiences. In other words, once neural states achieve the private status of subjective consciousness, whether they are mental images of the outside world, interoceptive states, or affects, they can never be objectively observed or experienced by someone else.

The Neurobiology versus the Subjectivity of Qualia

One implication of our analysis is that the explanation of the physical or neural basis of qualia is *not identical to* the explanation of their subjectivity. Indeed, we conclude that a major problem of past explanations of the gaps and the perplexing basis of qualia was their tendency to conflate these two issues, whereas we will show that the explanations for qualia and for their subjectivity, while overlapping, can be dissociated. Here is why.

We have established that the special neurobiological basis of *qualia* must have multiple causes—because qualia can be created by anatomically diverse and widespread brain structures—and we have also shown that *subjectivity* has multiple causes, from the "interiority" of life, from allo-irreducibility, and from auto-irreducibility, all of which contribute to the philosophical differences between the subjective experience of "being" a brain and the objective observation of that brain. But notice that the causes of qualia just listed are *not* the same as the causes of subjectivity just listed. So while the neurobiological causes of qualia, and the neurobiological-plus-philosophical causes of subjectivity, are interrelated, they are not identical.

Thus our explanation of the nature of qualia on the one hand as coming from unique neurobiological features of brains (box 6.3), and the reasons for their unique subjectivity on the other (fig. 8.2, and the embodiment of life), *are not the same explanations*. This is a critical point that appears to have been overlooked. Now we see that the neurobiology of qualia (again, box 6.3) does not fully explain qualia's subjectivity (again, fig 8.2), and the neurobiology of subjectivity does not fully explain the creation of qualia. Rather, it is the *combination* of the unique neurobiology of qualia and the unique features of subjectivity that explains the unique subjectivity of qualia.

Thus it appears that we are really dealing with two questions that are normally treated as one: qualia are *both* neurobiologically unique *and* exclusively first person (subjective), and the reasons for these two aspects of qualia are *different*. The uniqueness of qualia addresses the question "how are qualia generated?" which is explained by the special features of consciousness; and the subjectivity of qualia addresses the question "how can qualia be exclusively first person?" which is explained by the relationship between qualia and life, as well as the auto- and allo-irreducibilities. Therefore, trying to answer the question "how do neurons create subjective qualia?" is misguided because it seeks one answer when two are needed.

The Character of Experience

Now we can answer another question about the subjectivity of qualia: why do brain states "feel" the particular way that they do? David Chalmers called this the question of the *character of conscious experiences*: the perplexing question of why "red" subjectively feels exactly and uniquely the way red does. Or why does the activation of the auditory pathway lead to subjectively heard sounds? Isn't that beyond scientific explanation? Chalmers states it thus:

> Why do individual experiences have their particular nature? When I open my eyes and look around my office, why do I have *this* sort of experience? At a more basic level, why is seeing red like *this*, rather than like *that*! It seems conceivable that when looking at red things, one might have had the sort of color experiences that one in fact has when

looking at blue things. Why is the experience one way rather than the other? Why, for that matter, do we experience the reddish sensation that we do, rather than some entirely different sensation, like the sound of a trumpet?[11]

Our answer to this question is that once we understand that life processes "build in" the *potential* for subjective experience and that qualia are unique system features of the living brain that require the special features, then the reason why there are differences between the subjective feel of "red," the note "C-sharp," and feeling happy or sad becomes clear. We know that the neural pathways of color processing, those of sound processing, and those of affect are enormously different. They evolved to process different kinds of sensory stimuli and to direct different types of responses. Therefore they should not, and indeed *could not*, "feel" the same. It should come as no surprise, then, that the qualia these neural architectures create differ substantially among them. In other words, the *qualitative differentiation of neural states lies in the neural states themselves*. Qualia are integrated features of those states that are in turn based on the general and special features of consciousness (fig. 8.1) that create the unique neurobiological processes of qualia.

Thus "red" is *both* neurobiologically unique *and* subjective, with no explanatory gaps between the neurobiological creation of the feeling and its subjectivity. The supposed mystery of why red feels red is not solved by some *direct neurobiological connection* between the color-coded cones of the retina, or the "color areas" of the brain, and the subjective feeling of "red." Rather, the solution is that the feeling of "red" is *both* neurobiologically unique *and* subjective; and to explain how the neurons create the "feeling of red," we must consider *all* the contributing factors. These factors include life's contribution to subjectivity, and brains that are sufficiently complex (with the special features) and appropriately designed (with the general and special features) to create various subjective qualia (within the boundaries imposed by the allo- and auto-irreducibilities). For qualia to exist—and for red to feel "red"—all these factors must be present.

Why Is There Consciousness at All?

Finally, Chalmers wonders about what he calls the "hard problem" of consciousness: why and how should neural physical processes give rise to "experience" at all? Why don't all the brain functions that are associated with subjective experience and sensations go on "in the dark"?

> This further question is the key question in the problem of consciousness. Why doesn't all this information-processing go on "in the dark," free of any inner feel? Why is it that when electromagnetic waveforms impinge on a retina and are discriminated and categorized by a visual system, this discrimination and categorization is experienced as a sensation of vivid red? We know that conscious experience *does* arise when these functions are performed, but the very fact that it arises is the central mystery.[12]

Many writers on the subject of consciousness consider this the quintessential "hard problem of consciousness." We are not in a position to argue all the philosophical nuances and opinions regarding the hard problem, and we refer the reader to a number of in-depth discussions on the subject.[13] But here we offer our view on how neurobiological naturalism can demystify the "why consciousness?" aspect of the hard problem.

First, from the standpoint of neurobiology, wondering why the neural processes that create consciousness don't go on in the dark without consciousness is a bit like asking Ernst Mayr *why* the biological processes that create life don't go on without the integrated system feature of life.[14] Or like asking why the millions of multiplying embryonic cells that develop into a human body don't do so without development. These questions seem tautological at best and absurd at worst. Life is a naturally created and integrated system feature of atoms, molecules, tissues and organs, and so on. But life does not have a single cause, nor can life be simply reducible to its parts. In a similar way, sensory consciousness is ultimately a natural feature first of life, then of reflexes, and finally of complex brains. In this regard, why consciousness exists is no more mysterious than why life or embryonic development exists, but with the added feature of subjectivity. And we have explained the subjectivity part as stemming from embodiment and the ontological irreducibilities (fig. 8.2).

Another issue requires explanation. Chalmers distinguished what he called the "easy problems" from the hard problem of consciousness. The easy problems, according to his account, such as the ability to discriminate, categorize, and react to environmental stimuli, or to integrate information by a cognitive system, "are straightforwardly vulnerable to explanation in terms of computational or neural mechanisms,"[15] whereas the problem of experience—the hard problem—is not thus explainable. The line of reasoning here appears to be that if we can adequately explain so many brain functions and behaviors that can occur in the dark, via computational, cognitive, and even neural mechanisms, then how can we account for how brains create consciousness and qualia that occur "in the light," namely, experience? It is the latter aspect that appears to be left out of these accounts.

But our theory of neurobiological naturalism argues that animal experience is fundamentally and inextricably built on the foundation of life. Therefore, according our account, we must distinguish purely computational mechanisms—for example, those of computers or any other known nonliving computational device—as well as the *cognitive theories of consciousness* that likewise center on information processing, from theories that invoke the biological and neural properties of a living brain.[16] We hypothesize that experience and qualia are living processes that cannot be explained solely by a nonbiological computation, and our view of the hard problem begins and rests on the essential role that biology plays in making animal experience and qualia possible (plate 10).

That said, we should note that fairly complex reflexes and core brain functions—what Chalmers calls "easy problems"—do indeed operate in the dark. These nonconscious neural functions are living, so life, while necessary, is not sufficient for consciousness. What else is required? Again, the addition of the special features of consciousness (box 6.3) distinguishes nonconscious neural processing from consciousness, and the special features enable the in-the-dark reflexes and core brain functions to come into the light.

Therefore the more scientific question is not why consciousness doesn't go on in the dark but rather how conscious, subjective feeling is created by neurons and brains in the first place—and we offer a naturalized answer to that question, though, as we have demonstrated, it requires a host of variables to explain it. If our hypothesis is correct, the combination of life and reflexes, the special features, and auto- and allo-ontological irreducibilities can account for how both subjectivity and the unique phenomenon of consciousness are naturally created. Note that there is no *single* reason why and how this occurs. Rather, it occurs because many identifiable features across multiple, integrated, hierarchical biological levels, starting with life, collectively make consciousness possible.

In conclusion, we propose that the fundamental reasons that consciousness and the nature of subjective experience remain a mystery to many investigators is that they fail to distinguish between three related but different issues.

First, consciousness is incredibly diverse. This is true within brains—including different regions for mental images versus affects, or smell versus vision; and the diversity among the NSFCs such as qualia versus referral versus unity. It is also diverse across animals with very different brains—vertebrates, cephalopods, and arthropods. All this diversity means that consciousness and subjective experience cannot be explained by or reduced to a discrete group of biological or neurobiological factors; there can be no single consciousness "fundamental force" or process, or even a "core-conscious neurocircuit" shared by all. No "single bullet" theory. Rather, consciousness comes from the multifactorial combination of life and numerous unique neurobiological structures and processes.

Second, based on the first proposition, since the neurobiology of subjectivity is the result of so many diverse factors, and comes in so many forms and manifestations—including their diverse experiential aspects—philosophically treating qualia as "one thing" only partially and very broadly matches up with what we know about the diverse neurobiology and diverse qualities of subjective experience. This further suggests that "localizing" such a broad experiential feature as qualia in the brain is impossible, as is ascribing diverse qualia to just one kind of neural mechanism.

Third, once brains reach the level of neural hierarchical complexity and specialization that enables subjective experience, there arise the philosophical problems posed by the third-person observation of brains versus the very different aspects of subjective feeling—such as the "character" of experience. But subjective experience will always have different features than third-person explanations of subjectivity or brain processes. This is what Searle meant when he argued for the ontological subjectivity of consciousness.[17] But this does not mean that the neurobiology of consciousness cannot be scientifically explained. It just means that we must not conflate the neurobiological and the philosophical problems.

We propose that these three aspects of the aptly named "hard problem" are related but different problems, that they have different solutions, and that they cannot be reduced one to the other. Progress can only be made by separating these aspects.

Glossary

adaptation A goal-directed function (or a goal-meeting structure) in a living organism, as evolved by natural selection. See chap. 6.

affects, affective consciousness, affective states Qualitative feeling states that have a valence such as being positive or negative. Emotions are affects, and so are moods. They are global in that they involve the entire organism. They are one of the main categories of conscious experience, the other being mental images.

allo-ontological irreducibility See **auto- and allo-ontological irreducibility**.

amphioxus A fish-like invertebrate related to the vertebrates. Amphioxus is a member of the Cephalochordata and is the invertebrate thought to resemble most closely the ancestor of vertebrates. See fig. 3.5C.

arthropods The big phylum of jointed-leg animals consisting of insects and crustaceans (crabs, shrimps, barnacles, etc.), spiders and their relatives (scorpions, horseshoe crabs, etc.), and myriapods (millipedes, centipedes, and relatives). Trilobites and many other extinct arthropods are shown in fig. 7.3.

associative learning Learning an association between two stimuli or between a behavioral response and a stimulus. The two kinds are *classical conditioning* and **operant conditioning**. See chap. 4.

attention (selective attention) Focusing on important stimuli, filtering out unimportant stimuli, and shifting focus from one important stimulus to the next.

auto- and allo-ontological irreducibility *Auto-ontological irreducibility* means one's subjective consciousness never experiences, or "refers to," or is objectively aware of, the neurons that create experience (*auto* means "self"). *Allo-ontological irreducibility* means that an outside observer has no access to a subject's conscious experience (*allo* means "other [person]"). Both are key for explaining subjectivity. See fig. 8.2.

basal ganglia A region in the base of the vertebrate forebrain, which consists of the caudate nucleus and putamen, as well as the globus pallidus (pallidum) and nucleus accumbens. See figs. 4.1–4.2. Basal ganglia choose among a number of different motor programs to determine what behavioral actions are ordered. They have a "premotor" function, and also are involved in reward and motivation (nucleus accumbens).

bilaterian animals Animals whose bodies have right and left sides, mostly in mirror image; a front end with a mouth; a back end with an anus; and a through gut. Earthworms, insects, squids, and vertebrates are bilaterians. Bilaterians include all the multicellular animals except for sponges, jellyfish and their relatives, sea gooseberries (ctenophores), and obscure little flat animals called placozoans.

biological naturalism John Searle's theory that consciousness is real yet is entirely based on neurobiology. See chap. 1.

brain stem Three continuous parts of the lower brain in vertebrates—the medulla, pons, and midbrain—that connect the spinal cord with the higher brain structures. See fig. 4.1.

Cambrian explosion The rapid diversification of animals in the Cambrian Period (541 to 485 million years ago). Most of this diversification occurred in the first half of the Cambrian, from about 541 to 518 million years ago. See chap. 7.

Cambrian Period Time in Earth history from about 541 to 485 million years ago. Preceded by the Ediacaran (of the late Precambrian) and followed by the Ordovician.

camera eye Eye with a single focusing lens that forms a detailed image of the visual world on the retina. Vertebrates and cephalopods have such eyes, whereas arthropods form images with compound eyes instead (see **compound eye of arthropods**).

central nervous system (CNS) The brain and spinal nerve cord, as opposed to the nerves that run through the body and attach to the CNS. These nerves are located in the peripheral nervous system (PNS).

central pattern generator (CPG) A network of neurons that produces rhythmic, patterned outputs. Sensory input can modify its rhythm but does not cause the rhythm. CPGs underlie most motor programs and motor patterns, such as the repetitive movements of breathing, locomoting, and feeding. Even humans have many CPGs in the brain stem and basal forebrain (breathing centers, swallowing center, centers that influence the heartbeat, etc.).

cephalopods The most active of the mollusks: today's squids, cuttlefish, octopuses, and nautilus. The extinct groups of cephalopods, dating back to the Cambrian, also had the characteristic tentacles, large eyes, and locomotion by jet propulsion. See chap. 5.

cerebral cortex The uppermost (rostral) part of the forebrain in mammals, the cerebral cortex carries out the highest brain functions. It is the outer, gray layer of the mammalian cerebrum. It represents the *dorsal pallium* of other vertebrates. That is, all other vertebrates have a cerebrum (and dorsal pallium), but not an expanded cerebral cortex. See figs. 3.1–3.2.

chemoreceptors See **receptors**.

compound eye of arthropods Convex eye made up of many repeating eye units called *ommatidia*, each with its own lens and visual cells, and each acting as a separate visual receptor. A composite image of the visual field is built from all the ommatidia. The compound eye has lower resolution than the camera eyes of vertebrates and cephalopods but is great for detecting movements and provides an extremely wide visual field.

constitutive hierarchy In a constitutive hierarchy, all higher levels are physically composed of the elements at lower levels. For example, a person is made of organs that are in turn made of tissues that are made of cells, etc. See fig. 6.1.

core brain Primitive brain regions before consciousness, especially in the brain stem and diencephalon of vertebrates. See Lacalli 2008 and fig. 6.4.

deuterostomes The second main group of bilaterian animals other than the **protostomes**. The deuterostomes consist of vertebrates as well as some invertebrates: amphioxus, tunicates, and echinoderms (sea urchins, sea stars, sand dollars, sea lilies, sea cucumbers), plus acorn worms and their kin. Scientists originally recognized Deuterostomia because the first body opening to form in the embryo becomes the anus in most members, and gene analysis has now confirmed deuterostomes as a true, natural group. See fig. 7.5.

diencephalon A part of the vertebrate forebrain, between the midbrain and the telencephalon or cerebrum. Its main parts are the thalamus and hypothalamus. See figs. 3.1 and 4.1.

distance senses The exteroceptive senses that detect signals coming from far away: vision, hearing, smell, etc. Touch on the skin may or may not be included: feeling the wind coming from afar would be a distance sense, but feeling a nearby object that directly touches the skin might not. They do not include sensing taste on the tongue, the inner-body senses, or proprioception.

ectodermal placodes Collections of surface cells on the embryos of all vertebrates that develop into important vertebrate features: lens of the eye, smell receptors, some of the sensory neurons of the head, and the anterior part of the pituitary gland. See fig. 3.7.

embodiment Having a body, a key part of being a living organism. Embodiment entails a boundary, such as a cell membrane or skin that separates the organism's interior from the external environment.

explanatory gap The philosopher Joseph Levine coined this term to refer to the difficulty of fully explaining, equating, or reducing subjective experience to brain functions.

exteroception, exteroceptive consciousness Sensing stimuli that come from the external world (light, sound, touch, etc.), and the subjective experience of such stimuli as mapped mental images. Some investigators consider exteroception to be the prototypical form of consciousness.

forebrain In the vertebrates, the forebrain is the telencephalon and diencephalon (figs. 4.1 and 3.1), including the cerebrum and thalamus.

ganglion In certain invertebrates, a ganglion is an expanded mass of neurons and neuronal processes along the main nerve cord that forms the central nervous system. In the head region of such invertebrates, enlarged ganglia can form the brain. This is the case for the ganglionic brains of arthropods and cephalopod mollusks. In the vertebrates, by contrast, a ganglion is not a brain structure but a group of neuron cell bodies along a nerve out in the peripheral nervous system. The *basal ganglia* of vertebrates are misnamed because they are inside the brain. *Ganglion* is a Greek word meaning "knot on a string" or "tumor along a tendon."

gastropods Snails, slugs, and limpets. The most diverse taxonomic group of mollusks. Some gastropods mentioned in this book are the sea hare *Aplysia*, sea slug *Pleurobranchaea*, and snails *Lymnaea* and *Helix*.

general biological features The features of living organisms that distinguish life from nonlife. They are necessary but not sufficient for the creation of consciousness. They include life itself, embodiment, system and process, hierarchical functions, teleonomy, and adaptation. See box 6.1.

global Pertaining to the whole world, whole system, or whole body. Mostly used in the "whole-body" sense in this book.

global operant conditioning or learning Learning a brand-new behavior that applies to the whole body (chap. 4). See also **operant conditioning**.

habituation (learning by) A simple kind of learning by which an organism responds less and less to a stimulus as the stimulus is repeated over and over.

hard problem of consciousness The philosopher David Chalmers's term for the difficulty of explaining all aspects of subjective experience, even why subjective experience exists, on the basis of the physical properties and functions of the brain. It is related to Levine's **explanatory gap**. See chap. 8.

hierarchy An organized system composed of parts that are arranged in some sort of graded series of levels from low to high.

homeostasis The adjustment processes by which the body maintains a constant, optimal internal state even when perturbed or threatened (*homeo*, "constant"; *stasis*, "standing, state").

hypothalamus Part of the forebrain, the hypothalamus is the main visceral control center of the vertebrate brain. It is also a center for affects. See figs. 2.3 and 4.1.

images See **mental images**.

interoception, interoceptive consciousness Sensing of stimuli that come from inside the body, especially from the internal, visceral organs; and the associated conscious awareness of these stimuli. Examples are feeling a full stomach or bladder, nausea, and sore throat. Along with exteroceptive and affective consciousness, interoceptive is one of the three basic types of consciousness. It is intermediate between the other two types because it has mapped mental images but easily invokes affects. It also includes states such as hunger, thirst, fatigue, etc. Some classifications include the skin sensations of pain, itch, and tickle as forms of interoceptive awareness. See chap. 2.

isomorphic representations, isomorphism Neural representations that are organized according to point-to-point "maps" of bodily sensory fields or of the outside world (e.g., retinotopic mapping for vision, somatotopic for touch, tonotopic for audition), as well as nonspatially organized chemotopic representations (smell). See fig. 2.2 (plate 1) and **topographic mapping**.

judgment bias test A task that is designed to tell, objectively, if an animal experiences affects. After the animal is trained, by reward or nonreward or punishment, to distinguish between two stimuli, an ambiguous (intermediate) stimulus is presented. The speed and frequency of the animal's response to the ambiguous stimulus indicate whether it judges this intermediate stimulus to be positive or negative. For example, responding more quickly after a string of previous rewards indicates positive affect. See chap. 5 and fig. 5.3.

mental causation The capacity for personal mental events such as thoughts and feelings to have effects on the physical world, including on one's own brain and body.

mental images, image-based consciousness The mapped simulations of consciousness. We use the term *mental image* to refer to exteroceptive consciousness that creates referred, unified, causal, and qualitative mental representations. All exteroceptive images are in some sense maps. Interoceptive images of internal organs are included (pain in stomach). *Mental imagery* is not related to human "imagery," "imaginings" or "imagination." See fig. 2.4 (plate 2).

mental unity The capacity for the diversified and divisible brain to subjectively experience a unified field of awareness. It is one of the explanatory gaps and one of the neuro-ontologically subjective features of consciousness. See box 2.1.

modality, sensory modality A kind (mode) of sense, such as vision, hearing, smell, or touch. A multimodal mental image is one built from multiple senses.

neural correlates of consciousness (NCCs) According to David Chalmers (2000), the minimal set of neural structures or conditions that is sufficient for consciousness to occur. Many researchers seek these correlates in hope of examining them further to understand the neural basis of conscious experience. Sascha Fink (2016) strengthened the concept to make it easier to test whether the proposed correlates are correct: Fink defines NCCs as a necessary set of merely sufficient neural features. NCC is related to our **special neurobiological features of consciousness**.

neural crest An embryonic tissue found only in the vertebrates, whose cells give rise to nearly all the sensory neurons plus a wide variety of other structures (skeleton of the face, pigment cells, tooth dentin, core of the adrenal gland, etc.). See fig. 3.7.

neurobiological naturalism The theory we advocate in this book and elsewhere, adapted from Searle's **biological naturalism**. Neurobiological naturalism proposes that consciousness and subjectivity can naturally be explained by life processes plus the unique neurobiological features of evolved brains (box 6.3).

neuromodulator Chemical substance released by a neuron to influence multiple other neurons. Contrast with a **neurotransmitter**, which influences a single neuron. Examples include dopamine, serotonin, and norepinephrine. Neuromodulators are often associated with affective feelings.

neuron A nerve cell. See fig. 2.1.

neuro-ontologically subjective features of consciousness (NSFCs) The four properties of consciousness that are subjective and fall within the explanatory gaps. They are *referral, mental unity, mental causation,* and *qualia.* See box 2.1.

neurotransmitter Chemical substance released from a neuron at a synapse to influence the next neuron. Examples include acetylcholine, glutamate, gamma-aminobutyric acid (GABA), and glycine.

nociception The response of special receptors (*nociceptors*) to a "noxious stimulus," defined as a sensory stimulus that damages or threatens to damage tissues, such as a harmful mechanical, thermal, or chemical stimulus. One of the "touch" senses. Nociception leads to reflexive or even complex behavioral responses by the body, but not necessarily to conscious pain. See **pain**.

novel feature *Novel* means something never, ever, existed before, as opposed to *new*, which can mean "new for now."

objective Demonstrable to all, often by measurement, and independent of personal considerations; existing externally to the mind. Observable by any "third person." Opposite of **subjective**.

Onychophora The velvet worm phylum. Close relatives of the arthropods. See fig. 7.3.

operant conditioning or learning Learning from experience to approach a rewarding stimulus or to avoid a punishing one. See chap. 4.

pain The conscious experience of nociceptive stimuli or unpleasant affective experiences. See chap. 4.

pallium Part of the cerebrum of the vertebrate forebrain. Four subparts: *dorsal pallium* is the cerebral cortex in mammals, *lateral pallium* processes smell inputs, *medial pallium* is the hippocampus for forming and recalling memories, and *ventral pallium* is part of the amygdala, an emotional and fear center. See fig. 4.1.

phenomenal consciousness See **primary consciousness.**

phylum Group of related organisms that share the same basic body plan. About thirty-five animal phyla exist. The major phyla are the chordates (vertebrates, amphioxus, and tunicates), arthropods, mollusks, annelid worms, nematode roundworms, flatworms, echinoderms (sea stars, sand dollars, sea lilies, sea cucumbers), cnidarians (jellyfish and their kin), and sponges.

placodes See **ectodermal placodes.**

primary consciousness The basic ability to have subjective experiences including exteroceptive, interoceptive, or affective experiences. Primary consciousness includes the capacity to have *any* conscious mental images or affects and "something it is like to be" (Nagel 1974). It is not reflective, nor is it higher consciousness or self-consciousness.

primordial emotions The emotions that are most closely related to the survival of the organism, mainly for inner-body homeostasis and driven by interoceptors: thirst, hunger, air hunger, the need for salt, and so on. See Denton 2006.

projicience An aspect of **referral**. The process whereby stimuli from the distance senses, though received on the body surface, are experienced or "referred" out to the external world. An example of projicience is when light stimuli received on one's retina are experienced as originating from a point away from the eye; i.e., from the light source.

proprioception Sensing the amount of stretch in the body's joints, muscles, tendons, and skin as the body moves. Proprioception thereby informs the brain how the body is moving through space and where the body is positioned at all times. It is related to the sense of balance and has both conscious and unconscious aspects. The word *proprioception* means "sensing one's own [movements]." See chap. 2.

protostomes The largest group of bilaterian animals. This invertebrate group does not include humans or the other vertebrates. Includes dozens of phyla, the main ones being the arthropods, roundworms (nematodes), mollusks, annelid worms, and flatworms (platyhelminthes). This group was originally named because the first body opening to form in the embryo of most members becomes the mouth (*proto-stome*, "first mouth"). Gene analysis has largely confirmed Protostomia as a true, natural group. Except for a few small phyla, all bilaterian animals are either protostomes or **deuterostomes**. See chap. 5 and fig. 7.5.

qualia (singular: quale) A philosophical term to denote experienced qualities, or the way things feel. The subjective feeling of fear or the smell of a rose or the sound of middle C are examples of qualia.

receptors, sensory receptors Endings of sensory neurons (or special cells innervated by sensory neurons) that receive a stimulus and convert it to an electrical signal that travels along the sensory neuron. Some receptors respond to physical forces like touch (mechanoreceptors), some to light (photoreceptors), some to smell or taste (chemoreceptors), some to damaging stimuli (nociceptors), and some to electrical fields (electroreceptors). See the end of the "stimulus" arrow in fig. 2.1B.

reciprocal (reentrant, recurrent) neural communication Back-and-forth communication between neurons, brain regions, or the levels of a neural hierarchy. Cross talk with feedback and feedforward aspects.

referral The feature of consciousness in which the subjective experience is "mentally projected" away from the brain. This includes **projicience** of the **distance senses**, as well as all other sensory and affective processing that is never subjectively experienced as being in the brain itself but is always referred elsewhere to the external world, body, or self. See also **auto- and allo-ontological irreducibility**.

reflex An involuntary nervous reaction to a stimulus. The operation of a reflex does not require consciousness. An example of a reflex is the knee jerk, where a tap on the knee produces an involuntary jerk of the leg. This reflex will occur even if the person is asleep or in a coma.

reflex arc The neural pathway that brings about a reflex action. A reflex arc has, at least, a sensory neuron and a motor neuron, but most also have one or more interneurons in between. See figs. 2.1B and 6.2.

roundworms Worms of the nematode phylum, including *Caenorhabditis elegans* (*C. elegans*), which has only 302 neurons and is used to study simple nervous systems.

sensitization (learning by) A simple kind of learning by which an organism responds more to a stimulus as the stimulus is repeated over and over. The opposite of **habituation**.

sensory consciousness See **primary consciousness**.

sensory memory A very short memory of a sensory perception, allowing the continuity of sensory experience from moment to moment (described in chap. 6).

somatotopic map Topographic map of the body regions, especially the outer body, in the central nervous system, for the touch senses. See fig. 2.2 (plate 1) and **isomorphism**.

special neurobiological features of consciousness The unique features of complex nervous systems that, when added to the general biological features, create consciousness. They include elaborate sensory organs, the inputs from many senses joining within a complex brain, and neural hierarchies that create mapped mental images and affective states (for a full list, see box 6.3). In our theory, their evolution during the Cambrian marked the transition from reflexes and core brain functions to consciousness.

subjective, subjectivity The individual evaluation or experience of a conscious being, from the first-person point of view. Opposite of objective.

synapse A communicating junction between two neurons (nerve cells). See fig. 2.1B.

synchronized oscillations The theory of synchronized oscillations posits that neurons from different brain regions are reciprocally "bound" together in consciousness by their firing patterns matching in time. Through such communication, separate brain regions responsible for different features of a single object can be unified in consciousness. They are also thought to bind the different senses together in the unified image. See chap. 6.

system, complex system A complex unit with many parts, in which the arrangements and interactions of the parts are important.

tectum, optic tectum This forms most of the roof of the vertebrate midbrain and is a center for processing visual information throughout the vertebrates, though less so in mammals. The tectum also processes the other modes of sensory information (except smell). It is a relatively large part of the brain in fish and amphibians, and we propose it is the seat of image-based consciousness in those animals. See fig. 3.1, and plates 3 and 4.

telencephalon The large, frontal part of the vertebrate brain. Telencephalon's largest part is its cerebrum, or cerebral hemispheres.

teleonomy According to Mayr (2004), a goal-directed process that owes its goal or directedness to a program, often a genetic program. A teleonomic system does not foresee the end point of its operation. Rather, the achievement of an end result is built into the operation of the teleonomic process. Natural selection produces biological systems that are teleonomic, so teleonomy and evolutionary **adaptation** are closely related concepts. Teleonomic processes are prominent in development, physiology, and behavior. See chap. 6.

thalamus A deep region in the vertebrate forebrain, the thalamus plays a major role in sensory processing. It receives most of the sensory information that has come into the nervous system and relays this information to the cerebral cortex or cerebral pallium for further processing. See figs. 2.3 and 4.1.

topographic mapping Organized, point-by-point spatial arrangement of neural pathways between two parts of the nervous system. Almost the same as **isomorphism** except that its spatial maps are most obviously present in the touch-sensory and visual pathways.

tunicates Strangely specialized cousins of the vertebrates, with an expanded pharynx region for filter feeding and a distinctive, tough coat (tunic). Best-known members are the sea squirts and salps. Scientific name is Urochordata. Their gene sequences suggest that tunicates are more closely related to vertebrates than is amphioxus, despite their weirdness. See fig. 3.5A.

unity of consciousness See **mental unity**.

unlimited associative learning (UAL) An advanced way of learning an association between a stimulus and a reward or punishment, and thus a proposed indicator that an animal has affective consciousness (Bronfman, Ginsburg, and Jablonka 2016). UAL entails the ability to learn the reward value of a *whole* new object rather than of just one feature of the object, *plus* the ability to learn all the steps in a complex response needed to obtain the reward (or avoid the punishment), *plus* the ability to follow up and keep learning more and more. UAL may be so sophisticated that it reflects a more advanced version of affective consciousness than the simplest version we seek here. See chap. 4.

valence The subjective value, positive (good) or negative (bad), of a stimulus, object, or event. A basis of **affective consciousness**.

valence neuron A neuron that encodes and signals positive or negative value. Scientists recognize a valence neuron because it only responds, electrically, to a reward stimulus or else to a punishing stimulus.

vertebrates The group of animals that possess a spinal column (backbone). The vertebrates include fish, amphibians, reptiles, birds, and mammals.

viscera, visceral organs Internal organs of the body, especially those inside the chest cavity (lungs, heart, esophagus), abdominal cavity (liver, stomach and intestines, pancreas, kidneys, spleen), and pelvic cavity (bladder, uterus, ovaries, vagina, rectum).

vision-first hypothesis of the evolution of consciousness Idea that the evolution of image-forming eyes during the Cambrian led to the first conscious images, produced by the brains of the first arthropods and vertebrates. Advocated by Andrew Parker (2009), Michael Trestman (2013), and Feinberg and Mallatt (2016a).

Notes

1 What Makes Consciousness "Mysterious"?

1. Life differs from nonliving nature: Mayr 2004.

2. Explanatory gap: Levine 1983.

3. Nagel 1974, 436.

4. Primary or phenomenal consciousness: C. Allen and Bekoff 2010; G. Edelman 1989; Revonsuo 2006, 2010.

5. Revonsuo 2006, 37.

6. Theory of biological naturalism: Searle 1992, 2007, 2008, 2016.

7. Searle 1997, 212. Searle called the ultimately subjective nature of consciousness "ontological subjectivity."

8. Feinberg 2012; Feinberg and Mallatt 2016a, 2016b.

9. We consider "core brain" in chapter 6. Named by Lacalli (2008), it consists of the brain regions that are necessary for basic survival but not for consciousness. In vertebrates the core brain comprises the brain stem plus parts of the diencephalon. See figure 6.4.

10. Qualia, or qualities: Chalmers 1995a, 1995b, 1996; Crick and Koch 2003; Dennett 1988; Jackson 1982; Kirk 1994; Levine 1983; Metzinger 2004; Revonsuo 2006, 2010; Tye 2000.

2 Approaching the Gaps: Images and Affects

1. Three domains of consciousness: Feinberg and Mallatt 2016a.

2. Image-based consciousness: G. Edelman 1992, 112.

3. Damasio 2000, 2010.

4. Affective consciousness: Cabanac 1996; Cabanac, Cabanac, and Parent 2009; Panksepp 1998, 2005.

5. We use the word "interoception" itself in the classical way, to mean receiving inner-body sensations, mainly from the visceral organs (Sherrington 1906), so that the phrase "interoceptive consciousness" consists of two separable words. By contrast, Ceunen, Vlaeyen, and Van Diest (2016) use "interoception" to include the subjective, conscious aspects of sensing the body state. They call this a modern, broader usage, but it is based entirely on mammals and their brain's cerebral cortex, which stops any investigation of interoceptive consciousness in other animals that lack a cerebral cortex, and would thereby block our goal in this book.

6. Topographic and isomorphic maps: Hodos and Butler 1997. Also see J. Kaas 1997.

7. To read more about these types of mapped sensory organization, see any of the major neurobiology textbooks, such as Brodal 2016 and Kandel et al. 2012. For smell processing and the smell map, see Shepherd 2007 and Courtiol and Wilson 2014, respectively.

8. Interoceptive consciousness as central: Craig 2010; Denton 2006; Vierck et al. 2013.

9. Sensing stretch in the lung airways: Nonomura et al. 2017.

10. Damasio et al. 2000.

11. Pain as interoceptive consciousness: Craig 2003a, 2003b.

12. Classes of pain: Giordano 2005; Peirs and Seal 2016.

13. Proprioception: Proske and Gandevia 2012.

14. Motor aspect of consciousness: Barron and Klein 2016a; Cruse and Schilling 2015; Godfrey-Smith 2016a, 27, 81; Llinás 2002; Merker 2007; Morsella and Reyes 2016.

15. Feinberg 2012; Feinberg and Mallatt 2016a, 2016b.

16. Grain problem: Lockwood 1993; Meehl 1966; Sellars 1963, 1965.

17. Mental causation: Dardis 2008; Heil and Mele 1993; J. Kim 1998; Revonsuo 2010; Searle 2008; Walter and Heckmann 2003. See also *Internet Encyclopedia of Philosophy*, s.v. "emergence," http://www.iep.utm.edu/emergence.

18. Qualia as the central mystery: Crick 1995; Crick and Koch 2003; G. Edelman 1989; Jackson 1982; Kirk 1994; Metzinger 2004; Revonsuo 2006, 2010; Tye 2000.

3 Naturalizing Vertebrate Consciousness: Mental Images

1. Theories that all mammals and birds are conscious include those of Århem et al. 2008; Boly et al. 2013; Butler 2008; Butler and Cotterill 2006; D. Edelman and Seth 2009; G. Edelman, Gally, and Baars 2011; Harnad 2016.

2. Theories that consciousness requires a cerebral cortex and corticothalamic interactions include those of Baars, Franklin, and Ramsoy 2013; Key 2014; Lau and Rosenthal 2011; Tononi and Koch 2015. We cite many more corticothalamic theories in Feinberg and Mallatt 2016a.

3. Injury to cortex interferes with human consciousness: see Feinberg 2009; Scholarpedia, s.v. "touch disorders," http://www.scholarpedia.org/article/Central_touch_disorders; Wikipedia, s.v. "cortical blindness," https://en.wikipedia.org/wiki/Cortical_blindness; Wikipedia, s.v. "cortical deafness," https://en.wikipedia.org/wiki/Cortical_deafness.

4. Fish awareness: A. Abbott 2015; Balcombe 2016; Bshary and Grutter 2006; Schumacher, de Perera, and von der Emde 2017.

5. For a more detailed review of isomorphic mapping in vertebrates, see Feinberg and Mallatt 2013.

6. Birds' pallium is comparable to that of mammals, but with different sensory areas: Dugas-Ford, Rowell, and Ragsdale 2012; Jarvis et al. 2013; Karten 2013. For more on birds and image-based consciousness, see Marzluff et al. 2010; Stephan, Wilkinson, and Huber 2012.

7. Optic tectum and isomorphic maps: Butler and Hodos 2005; Guirado and Davila 2009; Knudsen 2011; Manger 2009; Northmore 2011; Robertson et al. 2006; Saidel 2009; Saitoh, Ménard, and Grillner 2007; Stein and Meredith 1993. The advantage of multiple senses contributing to a sensory image is that the image is more accurate than if from only one sense (Schumacher, de Perera, and von der Emde 2017).

8. Feinberg and Mallatt 2013, 2016a. Also, Newport et al. (2016) showed that archerfish can learn individual human faces, implying these fish had a remembered mental image and thus had a mental image, period.

9. Functions of the optic tectum: Ben-Tov et al. 2015; Del Bene et al. 2010; Graham and Northmore 2007; Gruberg et al. 2006; Gutfreund 2012; Kardamakis, Pérez-Fernández, and Grillner 2016; Nevin et al. 2010; Preuss et al. 2014; Schuelert and Dicke 2005; Temizer et al. 2015.

10. Tectum of fish and amphibians is for object recognition and perception: Dicke and Roth 2009; Wullimann and Vernier 2009. Similarly, Bianco and Engert (2015) say that in larval zebrafish a subset of tectal neurons is for "perceptual recognition of prey" because such neurons selectively respond to several different visual features of prey items: size, speed of motion, and light or dark tone.

11. The same shift in brain sites would have occurred, independently, in the evolution of birds (fig. 3.2). For reptiles, however, the condition is confusing because their cerebrum does not show much sensory isomorphism (in this way resembling fish and amphibians) but is somewhat expanded nonetheless, and their tectum, with its isomorphic sensory representations, is relatively large. Thus it is hard to tell if reptile consciousness is tectal or cortical. Furthermore, reptile brains are relatively poorly studied. For a full discussion of birds and reptiles, see Feinberg and Mallatt 2016a, 118–128, esp. 126–127.

12. Complexity of optic tectum: de Arriba and Pombal 2007; Marín, González, and Smeets 1997; McHaffie et al. 2005; Meek 1981; Northmore 2011; Saidel 2009; Wilczynski and Northcutt 1983. Internal tectal circuits are starting to be worked out: Bianco and Engert 2015, 843–844; Kardamakis, Pérez-Fernández, and S. Grillner 2016.

13. This is called declarative memory, and Woodruff (2017) summarized the evidence that it involves the hippocampus (or hippocampus-equivalent) in bony fish. Here is some more information of interest on the pallium of the forebrain. It performs many higher functions and is probably the most complex part of the brain of all vertebrates (see Feinberg and Mallatt 2016a). It processes information about smell, of course, but also integrates a lot of other sensory information to signal voluntary movements and behaviors. In fish, amphibians, and reptiles, the nonsmell sensory input reaches the pallium in highly processed form but is not isomorphically mapped, so it is difficult to know what role the pallium may play in nonsmell sensory consciousness in these particular vertebrates (though we have related it to conscious memory; see Feinberg and Mallatt 2016a, 112–115). Recently, Suryanarayana et al. (2017) worked out the essential neural architecture of the pallium of the lamprey, a basal jawless fish, and found it to resemble that of the mammalian cerebral cortex in its fundamentals, though the lamprey version seems much simpler. In fact, to us the lamprey pallium seems too simple, without many differences among its subregions, and with much of the integration being performed by the output neurons ("pyramidal tract cells"), plus just a few types of interneurons (see the simple circuit diagram in Suryanarayana et al. 2017). Lamprey behaviors are not simple, as they range from tracking their prey, to long-distance migrations, to mating rituals and nest building (Hardisty 1979; Hume et al. 2013; Swink 2003), so it seems puzzling that such simple pallial circuitry could cause or influence these behaviors. However, the authors of the study were only working out the *basic* circuitry, and future researchers may still discover many neural complexities. Additionally in lampreys, as in other vertebrates, the pallium communicates with many other parts of the brain (Northcutt and Wicht 1997), so in that respect it must be a complex hub of inputs, outputs, and neural processing.

14. See Feinberg and Mallatt 2016a, 115.

15. Reticular formation, etc.: Lee and Dan 2012.

16. Selective attention and consciousness: Marchetti 2014; Tsuchiya and van Boxtel 2013.

17. Lamprey tectum and attention: Kardamakis, Pérez-Fernández, and Grillner 2016.

18. Woodruff 2017; Ben-Tov et al. 2015.

19. Feinberg and Mallatt 2013, 2016a, 2016c.

20. Lampreys as predators: Swink 2003.

21. Lampreys as nest builders: Hume et al. 2013.

22. For evidence that fish discriminate among qualia, see Schumacher, de Perera, and von der Emde 2017.

23. Number of neurons in a conscious sensory hierarchy: Feinberg and Mallatt 2016a, 178.

24. Lacalli 2008, 2013, 2015, 2018. Lacalli's focus is more on amphioxus than tunicates; for tunicate larvae, what we know of CNS structure comes mainly from the laboratory of Ian Meinertzhagen. See, e.g., Ryan et al. 2016.

25. Lacalli 2018.

26. Vision spurred the evolution of image-based consciousness: Feinberg and Mallatt 2016a, 81–85; Lamb 2013; Parker 2009.

27. Ectodermal placodes and neural crest: Gans and Northcutt 1983; B. Hall 2008; Northcutt 2005; Schlosser 2014.

4 Naturalizing Vertebrate Consciousness: Affects

1. Affects can be divided into emotions and moods: Bethell 2015; Seth 2009a; Wikipedia, s.v. "mood (psychology)," https://en.wikipedia.org/wiki/Mood_(psychology). Emotions are shorter lasting than moods, more intense, and more likely to be signaled by a specific stimulus; an example

is a positive, pleasurable feeling. Moods, by contrast, are in the background and last longer; for example, a state of optimism or irritation.

2. Feinberg and Mallatt 2016a, chap. 8.

3. Operant learning: Brembs 2003a, 2003b; Perry, Barron, and Cheng 2013.

4. Unlimited associative learning: Bronfman, Ginsburg, and Jablonka 2016.

5. Classical learning in sea anemones, jellyfish relatives that have simple nervous systems: Haralson, Groff, and Haralson 1975. In roundworms, arthropods, vertebrates, snails, etc.: Perry, Barron, and Cheng 2013.

6. Feinberg and Mallatt 2016a, table 8.3.

7. Feinberg and Mallatt 2016a.

8. Cortex for emotions: Barrett et al. 2007; Berlin 2013; Craig 2010; LeDoux 2012; LeDoux and Brown 2017; Rolls 2014. Subcortical emotions: Damasio 2010; Damasio, Damasio, and Tranel 2012; Denton 2006; O'Connell and Hofmann 2011; Panksepp 1998, 2016; Solms 2013.

9. Strong emotions without cortex: Aleman and Merker 2014; Berridge and Kringelbach 2015, 651; Merker 2007; Panksepp et al. 1994.

10. Merker 2007, 79.

11. Aleman and Merker 2014.

12. Deep brain stimulation: Panksepp 2015.

13. Two parts of medial forebrain bundle: Panksepp 2016.

14. Affective brain structures in all the vertebrates: Feinberg and Mallatt 2016a, chap. 8.

15. Medial forebrain bundle in all vertebrates: Butler and Hodos 2005; Pombal and Puelles 1999.

16. Neuromodulator communication: Arbib and Fellous 2004, 558; Hu 2016; Namburi et al. 2016; Perry and Barron 2013; Zeman 2001.

17. Valence neurons: Betley et al. 2015; Beyeler et al. 2016; Felsenberg et al. 2017; Namburi et al. 2016. Hot spots: Berridge and Kringelbach 2015; Hu 2016; Namburi et al. 2016.

18. Thirst study: W. Allen et al. 2017. See also Gizowski and Bourque 2017.

19. Primordial emotions: Denton 2006.

20. Campos et al. 2018.

21. Affective centers: Feinberg and Mallatt 2016a, chap. 8. Amygdala for fear learning: Grewe et al. 2017. Arousal and adjusting intensity of emotions by part of anterior reticular formation (laterodorsal tegmental area): see LDT in fig. 4.2A and Brudzynski 2014; Rodríguez-Moldes et al. 2002; Ryczko et al. 2013. For more on different brain regions performing different affective functions, see Berridge and Kringelbach 2015; Damasio, Damasio, and Tranel 2012.

22. For more on these functions of the nucleus accumbens, basal ganglia, hypothalamus, and periaqueductal gray, see any neurobiology book, such as Brodal 2016.

23. Slug circuit: Gillette and Brown 2015. Simple, without hierarchical levels: Bronfman, Ginsburg, and Jablonka 2016, 6, 8.

24. Central pattern generators: Selverston 2010.

5 The Question of Invertebrate Consciousness

1. Some invertebrates have consciousness: Barron and Klein 2016; Cruse and Schilling 2016; D. Edelman and Seth 2009; S. Edelman, Moyal, and Fekete 2016; Elwood 2016; Godfrey-Smith 2016a; Klein and Barron 2016a, 2016b, 2016c; Mather and Carere 2016; Merker 2016; Montgomery 2015.

2. Arthropods have passed various tests for affective consciousness, namely, for operantly learned behaviors, behavioral trade-offs, frustra-

tion, and self-delivery of analgesics. Operantly learned behaviors: Abramson and Feinman 1990; Brembs 2003a; Kawai, Kono, and Sugimoto 2004; Kisch and Erber 1999; Tomina and Takahata 2010. Behavioral trade-offs: Elwood and Appel 2009; Herberholz and Marquart 2012; Stevenson and Schildberger 2013. Frustration: Pain 2009. Self-delivery of analgesics: Huber et al. 2011; Huston et al. 2013; Shohat-Ophir et al. 2012; Søvik and Barron 2013.

3. For a review of the affective tests on cephalopods, see Godfrey-Smith 2016a, 50–59. For operant and unlimited associative learning, see Andrews et al. 2013; Bronfman, Ginsburg, and Jablonka 2016; Cartron, Darmaillacq, and Dickel 2013; Crancher et al. 1972; Gutnick et al. 2011; Packard and Delafield-Butt 2014; Papini and Bitterman 1991. For behavioral trade-offs, see Anderson and Mather 2007; Mather and Kuba 2013.

4. Arthropods have the features of image-based consciousness: see Feinberg and Mallatt 2016a, table 9.2; Strausfeld 2013. For a fly brain having a representation of where the fly is heading, see S. Kim et al. 2017. For the somatosensory (touch) and taste pathways of flies being topographically mapped and separated by sensory modality, see Tsubouchi et al. 2017. Arthropod brains have small numbers of neurons: Feinberg and Mallatt 2016a, 181.

5. Neuron numbers, insects versus vertebrates: Chittka and Niven 2009; Wikipedia, s.v. "list of animals by number of neurons," http://en.wikipedia.org/wiki/List_of_animals_by number_of_neurons.

6. Neuron numbers in cephalopods: Hochner 2012.

7. Limited learning in nematode worms: Bronfman, Ginsburg, and Jablonka 2016; Perry, Barron, and Cheng 2013.

8. No image-based consciousness in nematode worms: Klein and Barron 2016a; see also Ardiel and Rankin 2010.

9. Nematodes' systematic search: Hills 2006. Nematodes are not conscious foragers: Klein and Barron 2016a.

10. Some snails may be close to consciousness: Bronfman, Ginsburg, and Jablonka 2016, 6–8. Second-order learning: Loy, Fernández, and

Acebes 2006. Blocking by competing cues: Prados et al. 2013. Snail eyes: Ziegler and Meyer-Rochow 2008. Only simple operant learning in gastropods: Brembs 2003a, 2003b.

11. Montgomery 2015; Godfrey-Smith 2016a. For more on cephalopod consciousness, see Darmaillacq, Dickel, and Mather 2014.

12. Cephalopod studies: D. Edelman and Seth 2009; Gutnick et al. 2011; Hochner 2013; Hochner, Shomrat, and Fiorito 2006; Mather 2012; Mather and Carere 2016; Mather and Kuba 2013.

13. For a discussion of play, both in general and in octopuses, see Burghardt 2005; Mather and Anderson 1999.

14. For more on the controversy over whether arthropods have consciousness, see Barron and Klein 2016a, 2016b; Klein and Barron 2016; and the debates in the journal *Animal Sentience*, including Adamo 2016a, 2016b; Cruse and Schilling 2016; Elwood 2016; Key 2016; Klein and Barron 2016b, 2016c; Mallatt and Feinberg 2016; Merker 2016; Shanahan 2016; Søvik and Perry 2016; Tye 2016.

15. Debates over whether arthropods are uniform: Tye 2016.

16. Fossil brains of early arthropods and near arthropods: Cong et al. 2014; Ma et al. 2012.

17. Bees and mental images: Fauria, Colborn, and Collett 2000.

18. One could argue that no, this is not consciousness but just implicit associative learning, a simpler phenomenon. Here is our response: the idea is that the two patterns together are visually complex, and then they unexpectedly become separated in time and space, adding even more complexity, so that only a remembered, detailed mental image could allow the bees to decipher the puzzle and find the food.

19. Bees and affects: Perry, Baciadonna, and Chittka 2016.

20. For more on the judgment bias test, see Bethell 2015.

21. Multiple realizability: Piccinini and Craver 2011, 301–302.

6 Creating Consciousness: The General and Special Features

1. Finding the common "neural correlates of consciousness," especially in the cerebral cortex, is the goal of many in the field of consciousness studies. Here is a sampling: Aru et al. 2012; Chalmers 2000; D. Edelman et al. 2005; Fink 2016; Hohwy 2007; Mormann and Koch 2007; Reggia 2013; Searle 2007; Seth 2009a; Seth, Baars, and Edelman 2005.

2. A cerebral cortex is not needed but lesser brain regions suffice: Barron and Klein 2016; Bronfman, Ginsburg, and Jablonka 2016; Ginsburg and Jablonka 2010; Merker 2007.

3. Life and subjectivity are personal: Thompson 2007.

4. Life and consciousness are processes: James 1904.

5. System and hierarchy theory: Ahl and Allen 1996; T. Allen and Starr 1982; Mayr 1982; Salthe 1985; Simon 1962, 1973. Constitutive hierarchies: Mayr 1982; see also Feinberg 2000.

6. Emergent properties: Campbell 1974; Clayton 2006; Mayr 1982; Morowitz 2004; Pattee 1970; Salthe 1985.

7. Emergence and consciousness: Beckermann, Flohr, and Kim 1992; Chalmers 2006; Craver 2007; Feinberg 2001, 2012; Feinberg and Mallatt 2016a; J. Kim 1992, 2006; Mallatt and Feinberg 2017; Searle 1992, 11; Van Gulick 2001.

8. Searle calls this view of emergence "emergent1." For his discussion of the philosophy of emergence with reference to consciousness, see Searle 1992, 112–126.

9. Teleonomy and adaptation: coined by the biochemist Jacques Monod, these definitions derive from Mayr 2004.

10. Nerve net in jellyfish and their kin: Bosch et al. 2017; Ruppert, Fox, and Barnes 2004.

11. Complex reflexes in a roundworm: from Mark Alkema and his laboratory; see Fang-Yen, Alkema, and Samuel 2015. See also Pirri et al. 2009.

12. Core brain in vertebrates and insects: Barron and Klein 2016, 7; Lacalli 2008; Merker 2007, 2016; Søvik and Perry 2016.

13. Core brain is for maintaining homeostasis, patterned locomotion, arousal, attention, and motivation: Brodal 2016; Grillner et al. 2005; Nieuwenhuys, Veening, and Van Domburg 1987; Parvizi and Damasio 2003.

14. The reticular formation projects widely: Brodal 2016.

15. Core brain in larval amphioxus: Lacalli 2018.

16. Arousal and attention in insects: Van Swinderen and Andretic 2011.

17. Arousal neurons influence locomotion in insects: De Bivort and Van Swinderen 2016.

18. The core brain in insects and motivation: Klein and Barron 2016a, 7.

19. Only image-based consciousness allows directed, planned movements through complex space in the absence of immediate stimuli: Barron and Klein 2016; Hills 2006; Merker 2007, 2016; Søvik and Perry 2016.

20. Special features: Feinberg and Mallatt 2016a.

21. Neuron types: Jabr 2012; Kandel et al. 2012; Strausfeld 2013; Underwood 2015; Yoshinaga and Nakajima 2017; Zeisel et al. 2015.

22. For information on the complex senses of vertebrates, arthropods, and cephalopods, see Kardong 2012; Kuba, Gutnick, and Hochner 2012; Mather 2012; Strausfeld 2013.

23. For more discussion of the special nature of neural hierarchies, see Feinberg 2011; Feinberg and Mallatt 2016a, chap. 2.

24. Reciprocal communication in conscious systems: Lamme 2006.

25. For a sample of the huge literature on reciprocal waves of oscillation binding neural information for consciousness, see Feinberg and Mallatt 2016a, 266n13, 268n28. Here we add Akam and Kullman 2014; Khodagholy, Gelinas, and Buzsáki 2017; Krebber et al. 2015; Min 2010; Paulk

et al. 2013; Randall, Whittington, and Cunningham 2011. Traditional ideas about the importance of gamma-frequency waves in this process are challenged in a recent review article by Koch et al. (2016).

26. Tononi 2011; Tononi and Koch 2015.

27. Multisensory convergence in brains of cephalopods and arthropods: Graindorge et al. 2006; Hochner 2013; Klein and Barron 2016a.

28. Prediction: Bronfman, Ginsburg, and Jablonka 2016; Clark 2013; Eshel et al. 2015; Gershman, Horvitz, and Tenenbaum 2015; Llinás 2002; Schultz 2015; Seth 2013.

29. Attention in vertebrates: see chap. 3 and Krauzlis, Lovejoy, and Zénon 2013.

30. Attention in insects: De Bivort and Van Swinderen 2016. See also Van Swinderen and Andretic 2011.

31. Long-term memory in insects: Comas, Petit, and Preat 2004; Tonoki and Davis 2015. In fish: Schluessel and Bleckmann 2005.

32. Types of memory: Koch 2004. See also Kaas, Stoeckel, and Goebel 2008; The Human Memory, "Sensory Memory," http://www.human-memory.net/types_sensory.html; BYU David O. McKay School of Education, "Cognition: Sensory Memory," http://byuipt.net/564/2013/08/23/cognition-sensory-memory.

33. Koch 2004.

7 The Evolution of Primary Consciousness and the Cambrian Hypothesis

1. Documentation for figure 7.1, plate 8: oldest bacterium-like fossils are 3.7 billion years old: Nutman et al. 2016; Schopf and Kudryavtsev 2012. A controversial study by Tashiro et al. (2017), however, says 3.95 billion, based on rock chemistry rather than cell fossils. First multicellular animals at 660–600 mya: Brocks et al. 2017. To deduce the time of the first animals with nervous systems, we used the physical evidence: the date of the oldest body fossils and trace fossils (trace fossils are

evidence of behavior, such as worm trails or scratch marks in the mud or sand of the ancient seafloor, since petrified). See Pecoits et al. 2012. The first arthropods are attested by scratch marks at 540 mya (Mángano and Buatois 2014), and first vertebrates by body fossils at 520 mya (Shu et al. 2003). Other investigators date these events by another method, using "molecular clocks" obtained from the gene sequences of modern animals, and they come up with older dates of origin for the basic animal groups. For a full discussion of this clock dating, see Erwin and Valentine 2013.

2. To be precise, radioactive dating says the whole Cambrian Period lasted from 541 to 485 million years ago. See Erwin and Valentine 2013. Besides us, other researchers who have proposed that consciousness arose during the Cambrian explosion are Barron and Klein 2016; Bronfman, Ginsburg, and Jablonka 2016; Godfrey-Smith 2016b; Packard and Delafield-Butt 2014; Trestman 2013; and Verschure 2016.

3. The evidence for a simple ancestral worm is given in Feinberg and Mallatt 2016a, 61–62. For new evidence supporting this claim, obtained using genes to build a tree of life, see Cannon et al. 2016.

4. Pecoits et al. (2012) report on these worm trails. See also Klein and Barron 2016a; Carbone and Narbonne 2014.

5. Few behavioral interactions in the Precambrian: Godfrey-Smith 2016b.

6. Predation as key to the Cambrian explosion: Erwin and Valentine 2013. Scavenging before predation: Godfrey-Smith 2016a, 37, after James Gehling, and Schiffbauer et al. 2016.

7. Distance senses: Lamb 2013; Parker 2009; Plotnick, Dornbos, and Chen 2010; Trestman 2013.

8. Evolution of cephalopod consciousness: Kröger, Vinther, and Fuchs 2011. See also Godfrey-Smith 2016a. Another kind of living cephalopod is the nautilus, and its brain and eyes are somewhat less developed than those of the octopuses, cuttlefish, and squid (Shigeno 2017). Perhaps the nautilus is not conscious?

9. We can tell exactly what these distance senses were because they are shared by today's lampreys, sharks, and bony fish. They included detailed vision, hearing, smell, and taste. The first fish also had electroreception (detecting objects in the water by sensing their electrical fields; see Bellono, Leitch, and Julius 2017) and lateral-line mechanoreceptors, which detect objects moving at a distance by sensing the vibrations those movements cause in seawater (Kardong 2012).

10. Camera eyes tell the most: Trestman 2013.

11. Smell first: Kohl 2013; Plotnick, Dornbos, and Chen 2010. Vision first, because no smell organ occurs in amphioxus or tunicates: Feinberg and Mallatt 2016a, 83–84; Kardong 2012; Vopalensky et al. 2012.

12. Lobopodians: Ortega-Hernández 2015.

13. Onychophoran eyes and brains: Homberg 2008; Mayer 2006; Schumann, Hering, and Mayer 2016; Strausfeld 2013, 384–391. These references also show that onychophoran worms have tentacles for smell and a good smell-processing region in their brain. Such facts suggest that smell perception played an earlier role in arthropod evolution than it did in vertebrate evolution.

14. Gastropods have a simple nervous system: Baxter and Byrne 2006; Ruppert, Fox, and Barnes 2004; Wikipedia, s.v. "nervous system of gastropods," https://en.wikipedia.org/wiki/Nervous_system_of_gastropods.

15. Affective and premotor centers in insects: Davis et al. 2014; Perry and Barron 2013; Søvik, Perry, and Barron 2015; Waddell 2013.

16. Note 1 of this chapter documents these dates and stages.

17. These ideas on the adaptive functions of consciousness come from the following sources. See Seth 2009b, where he summarizes literature on how consciousness integrates many other neural processes; how it simulates the world to allow better and more flexible behavioral interactions with that world (Baars 1988; Dehaene and Naccache 2001); and how conscious attention allows better learning and memory (G. Edelman 2003; Lamme and Roelfsema 2000). Consciousness is the best for making choices for actions (Denton 2006, 5; Jonkisz 2015; Keller 2014),

especially for selecting actions in complex environments (Damasio 2010, 61–62, 284). The optic tectum and affective brain structures of vertebrates, both of which are heavily involved in consciousness, send outputs to the basal ganglia, which then select motor programs for actions: de Arriba and Pombal 2007; McHaffie et al. 2005.

18. Consciousness replaces the need "to have myriads of neural programs to account for all the situations they encounter and adaptive responses": Ristau 2016, referring to a statement by the animal ethologist Donald Griffin.

19. Consciousness resolves inconsistencies between conflicting sensory inputs: Seth 2009b, after Morsella 2005.

20. Important items influence the motor program most: Andersen et al. 2015.

21. Importance ranking and affects: Cabanac 1996.

22. No spatial sense means no guided behaviors: Barron and Klein 2016; Merker 2007, 2016.

23. For another explanation of how consciousness allows behavioral flexibility, from studies that model neural complexity, see Seth 2009a, 53–54.

8 Naturalizing Subjectivity

1. This chapter on the important link between life and consciousness expands on an idea we presented in Feinberg 2012; Feinberg and Mallatt 2016a, 195–196; and Feinberg and Mallatt 2016b. For an extensive analysis of the life-consciousness link from a more philosophical perspective, see Thompson 2007. Terrence Deacon (2011) presents a complex theory of the relationship between mind and matter with reference to thermodynamics, emergence, and self-organizing systems; and Mayr (2004, 35) also noted that the philosophy of biology was essential to explaining consciousness and the mind.

2. Chalmers (1995a, 1996) proposed that consciousness should be taken as "fundamental" and is not explainable by "anything simpler."

3. Brodal 2016; Grillner et al. 2005; Nieuwenhuys, Veening, and Van Domburg 1987; Parvizi and Damasio 2003. For further references on the reticular formation and thalamus, see Feinberg and Mallatt 2016a, 269n46.

4. Crick and Koch 2003, 119. The "hard problem" mentioned in this quotation is the term coined by David Chalmers that addresses how and why the *physical* brain can create *subjective* experience.

5. Note again that we are not saying life *alone* is sufficient for qualia, only that life is *prerequisite* for animal qualia.

6. Feinberg 2012; Feinberg and Mallatt 2016a, 2016b.

7. Globus 1973, 1129.

8. Autocerebroscope: Feigl 1967, 14.

9. Glia cells: N. Abbott 2004; J. Hall 2011; Oberheim, Goldman, and Nedergaard 2012; Verkhratsky and Parpura 2014.

10. Allo-ontological irreducibility: Feinberg 2012; Feinberg and Mallatt 2016a, 2016b.

11. Chalmers 1996, 5.

12. Chalmers 1995a, 203.

13. More on the hard problem of consciousness: Bruiger 2017; Churchland 1996; Dennett 1991; McGinn 1991; Pigliucci 2013; Searle 2016; *Internet Encyclopedia of Philosophy*, s.v. "hard problem of consciousness," http://www.iep.utm.edu/hard-con; Scholarpedia, s.v. "hard problem of consciousness," http://www.scholarpedia.org/article/Hard_problem_of_consciousness.

14. Mayr 2004. For a philosophical perspective on this point, see Daniel C. Dennett, "Facing Backwards on the Problem of Consciousness," November 10, 1995, https://ase.tufts.edu/cogstud/dennett/papers/chalmers.htm.

15. More on the easy problems of consciousness: Chalmers 1995a.

16. Strictly computational and cognitive models fall short because they do not adequately account for the role of life in consciousness. For how computers differ from brains, see Sterling and Laughlin 2015. For information on current computational and cognitive models, see Baars and McGovern 1996; Block 2007; Dehaene 2014; Dretske 1995; Gennaro 2011; Metzinger 2004; Milkowski 2013; Piccinini 2015; Velmans and Schneider 2008; *Stanford Encyclopedia of Philosophy*, s.v. "consciousness," https://plato.stanford.edu/entries/consciousness; *Stanford Encyclopedia of Philosophy*, s.v. "computational theory of mind," https://plato.stanford.edu/entries/computational-mind.

17. Searle 1992.

References

Abbott, A. 2015. Clever fish. *Nature* 521:413–414.

Abbott, N. J. 2004. Evidence for bulk flow of brain interstitial fluid: Significance for physiology and pathology. *Neurochemistry International* 45 (4): 545–552.

Abramson, C. I., and R. D. Feinman. 1990. Lever-press conditioning in the crab. *Physiology and Behavior* 48 (2): 267–272.

Adamo, S. A. 2016a. Do insects feel pain? A question at the intersection of animal behaviour, philosophy and robotics. *Animal Behaviour* 118: 75–79.

Adamo, S. 2016b. Subjective experience in insects: Definitions and other difficulties. *Animal Sentience: An Interdisciplinary Journal on Animal Feeling* 1 (9): 15.

Ahl, V., and T. F. H. Allen. 1996. *Hierarchy Theory*. New York: Columbia University Press.

Akam, T., and D. M. Kullmann. 2014. Oscillatory multiplexing of population codes for selective communication in the mammalian brain. *Nature Reviews Neuroscience* 15 (2): 111–122.

Aleman, B., and B. Merker. 2014. Consciousness without cortex: A hydranencephaly family survey. *Acta Paediatrica* 103 (10): 1057–1065.

Allen, C., and M. Bekoff. 2010. Animal consciousness. In *The Blackwell Companion to Consciousness*, ed. M. Velmans and S. Schneider. Malden, MA: Blackwell.

Allen, T. F. H., and T. B. Starr. 1982. *Hierarchy: Perspectives for Ecological Complexity*. Chicago: University of Chicago Press.

Allen, W. E., L. A. DeNardo, M. Z. Chen, C. D. Liu, K. M. Loh, L. E. Fenno, et al. 2017. Thirst-associated preoptic neurons encode an aversive motivational drive. *Science* 357 (6356): 1149–1155.

Andersen, B. S., C. Jørgensen, S. Eliassen, and J. Giske. 2015. The proximate architecture for decision-making in fish. *Fish and Fisheries* 17:680–695.

Anderson, R. C., and J. A. Mather. 2007. The packaging problem: Bivalve prey selection and prey entry techniques of the octopus *Enteroctopus dofleini*. *Journal of Comparative Psychology* 121 (3): 300.

Andrews, P. L., A. S. Darmaillacq, N. Dennison, I. G. Gleadall, P. Hawkins, J. B. Messenger, et al. 2013. The identification and management of pain, suffering, and distress in cephalopods, including anaesthesia, analgesia, and humane killing. *Journal of Experimental Marine Biology and Ecology* 447:46–64.

Arbib, M. A., and J. M. Fellous. 2004. Emotions: From brain to robot. *Trends in Cognitive Sciences* 8 (12): 554–561.

Ardiel, E. L., and C. H. Rankin. 2010. An elegant mind: Learning and memory in *Caenorhabditis elegans*. *Learning and Memory (Cold Spring Harbor, NY)* 17 (4): 191–201.

Århem, P., B. I. B. Lindahl, P. R. Manger, and A. B. Butler. 2008. On the origin of consciousness—some amniote scenarios. In *Consciousness Transitions: Phylogenetic, Ontogenetic, and Physiological Aspects*, ed. H. Liljenstrom and P. Århem, 77–96. San Francisco: Elsevier.

Aru, J., T. Bachmann, W. Singer, and L. Melloni. 2012. Distilling the neural correlates of consciousness. *Neuroscience and Biobehavioral Reviews* 36 (2): 737–746.

Baars, B. J. 1988. *A Cognitive Theory of Consciousness*. New York: Cambridge University Press.

Baars, B. J., S. Franklin, and T. Z. Ramsoy. 2013. Global workspace dynamics: Cortical "binding and propagation" enables conscious contents. *Frontiers in Psychology* 4 (200).

Baars, B. J., and K. McGovern. 1996. Cognitive views of consciousness. In *The Science of Consciousness: Psychological, Neuropsychological, and Clinical Reviews*, ed. M. Velmans, 63–95. New York: Routledge.

Balcombe, J. 2016. *What a Fish Knows: The Inner Lives of Our Underwater Cousins*. New York: Macmillan.

Barrett, L. F., B. Mesquita, K. N. Ochsner, and J. J. Gross. 2007. The experience of emotion. *Annual Review of Psychology* 58:373.

Barron, A. B., and C. Klein. 2016. What insects can tell us about the origins of consciousness. *Proceedings of the National Academy of Sciences of the United States of America* 113 (18): 4900–4908.

Baxter, D. A., and J. H. Byrne. 2006. Feeding behavior of *Aplysia*: A model system for comparing cellular mechanisms of classical and operant conditioning. *Learning and Memory* 13 (6): 669–680.

Beckermann, A., H. Flohr, and J. Kim, eds. 1992. *Emergence or Reduction? Essays on the Prospects of Nonreductive Physicalism*. Berlin: Walter de Gruyter.

Bellono, N. W., D. B. Leitch, and D. Julius. 2017. Molecular basis of ancestral vertebrate electroreception. *Nature* 543 (7645): 391–396.

Ben-Tov, M., O. Donchin, O. Ben-Shahar, and R. Segev. 2015. Pop-out in visual search of moving targets in the archer fish. *Nature Communications* 6 (6476): 1–11.

Berlin, H. 2013. The brainstem begs the question: Petitio principii. *Neuro-psychoanalysis* 15 (1): 25–29.

Berridge, K. C., and M. L. Kringelbach. 2015. Pleasure systems in the brain. *Neuron* 86 (3): 646–664.

Bethell, E. J. 2015. A "how-to" guide for designing judgment bias studies to assess captive animal welfare. In *Advancing Zoo Animal Welfare Science and Policy: Selected Papers from the Detroit Zoological Society 3rd International Symposium* (November 2014). Supplement, *Journal of Applied Animal Welfare Science* 18 (S1): S18–S42.

Betley, J. N., S. Xu, Z. F. H. Cao, R. Gong, C. J. Magnus, Y. Yu, and S. M. Sternson. 2015. Neurons for hunger and thirst transmit a negative-valence teaching signal. *Nature* 521 (7551): 180–185.

Beyeler, A., P. Namburi, G. F. Glober, C. Simonnet, G. G. Calhoon, G. F. Conyers, et al. 2016. Divergent routing of positive and negative information from the amygdala during memory retrieval. *Neuron* 90 (2): 348–361.

Bianco, I. H., and F. Engert. 2015. Visuomotor transformations underlying hunting behavior in zebrafish. *Current Biology* 25 (7): 831–846.

Block, N. 2007. Consciousness, accessibility, and the mesh between psychology and neuroscience. *Behavioral and Brain Sciences* 30 (5–6): 481–499.

Boly, M., A. K. Seth, M. Wilke, P. Ingmundson, B. Baars, S. Laureys, et al. 2013. Consciousness in humans and non-human animals: Recent advances and future directions. *Frontiers in Psychology* 4 (625).

Bosch, T. C., A. Klimovich, T. Domazet-Lošo, S. Gründer, T. W. Holstein, G. Jékely, et al. 2017. Back to the basics: Cnidarians start to fire. *Trends in Neurosciences* 40:92–105.

Brembs, B. 2003a. Operant conditioning in invertebrates. *Current Opinion in Neurobiology* 13 (6): 710–717.

Brembs, B. 2003b. Operant reward learning in *Aplysia*. *Current Directions in Psychological Science* 12 (6): 218–221.

Brocks, J. J., A. J. Jarrett, E. Sirantoine, C. Hallmann, Y. Hoshino, and T. Liyanage. 2017. The rise of algae in Cryogenian oceans and the emergence of animals. *Nature* 548 (7669): 578–581.

Brodal, P. 2016. *The Central Nervous System: Structure and Function*. 5th ed. New York: Oxford University Press.

Bronfman, Z. Z., S. Ginsburg, and E. Jablonka. 2016. The transition to minimal consciousness through the evolution of associative learning. *Frontiers in Psychology*, December. https://doi.org/10.3389/fpsyg.2016.01954.

Brudzynski, S. M. 2014. The ascending mesolimbic cholinergic system—a specific division of the reticular activating system involved in the initiation of negative emotional states. *Journal of Molecular Neuroscience* 53 (3): 436–445.

Bruiger, D. 2017. Can science explain consciousness? *Philosophical Papers*. https://philpapers.org/archive/DANCSE-2.pdf.

Bshary, R., and A. S. Grutter. 2006. Image scoring and cooperation in a cleaner fish mutualism. *Nature* 441 (7096): 975–978.

Burghardt, G. M. 2005. *The Genesis of Animal Play: Testing the Limits*. Cambridge, MA: MIT Press.

Butler, A. B. 2008. Evolution of brains, cognition, and consciousness. *Brain Research Bulletin* 75 (2): 442–449.

Butler, A. B., and R. M. Cotterill. 2006. Mammalian and avian neuroanatomy and the question of consciousness in birds. *Biological Bulletin* 211 (2): 106–127.

Butler, A. B., and W. Hodos. 2005. *Comparative Vertebrate Neuroanatomy*. 2nd ed. Hoboken, NJ: Wiley Interscience.

Cabanac, M. 1996. On the origin of consciousness, a postulate and its corollary. *Neuroscience and Biobehavioral Reviews* 20 (1): 33–40.

Cabanac, M., A. J. Cabanac, and A. Parent. 2009. The emergence of consciousness in phylogeny. *Behavioural Brain Research* 198 (2): 267–272.

Campbell, D. T. 1974. Downward causation in hierarchically organized biological systems. In *Studies in the Philosophy of Biology*, ed. F. J. Ayala and T. Dobzhansky, 179–186. Berkeley: University of California Press.

Campos, C. A., A. J. Bowen, C. W. Roman, and R. D. Palmiter. 2018. Encoding of danger by parabrachial CGRP neurons. *Nature* 555: 617–622.

Cannon, J. T., B. C. Vellutini, J. Smith, F. Ronquist, U. Jondelius, and A. Hejnol. 2016. Xenacoelomorpha is the sister group to Nephrozoa. *Nature* 530 (7588): 89–93.

Carbone, C., and G. M. Narbonne. 2014. When life got smart: The evolution of behavioral complexity through the Ediacaran and early Cambrian of NW Canada. *Journal of Paleontology* 88 (2): 309–330.

Cartron, L., A. S. Darmaillacq, and L. Dickel. 2013. The "prawn-in-the-tube" procedure: What do cuttlefish learn and memorize? *Behavioural Brain Research* 240:29–32.

Ceunen, E., J. W. Vlaeyen, and I. Van Diest. 2016. On the origin of interoception. *Frontiers in Psychology* 7:743.

Chalmers, D. J. 1995a. Facing up to the problem of consciousness. *Journal of Consciousness Studies* 2:200–219.

Chalmers, D. J. 1995b. The puzzle of conscious experience. *Scientific American* 273 (6): 80–87.

Chalmers, D. J. 1996. *The Conscious Mind: In Search of a Fundamental Theory*. New York: Oxford University Press.

Chalmers, D. J. 2000. What is a neural correlate of consciousness? In *Neural Correlates of Consciousness: Empirical and Conceptual Questions*, ed. T. Metzinger, 17–40. Cambridge, MA: MIT Press.

Chalmers, D. J. 2006. Strong and weak emergence. In *The Re-emergence of Emergence*, ed. P. Clayton and P. Davies, 244–256. New York: Oxford University Press.

Chittka, L., and J. Niven. 2009. Are bigger brains better? *Current Biology* 19 (21): R995–R1008.

Churchland, P. M. 1996. The rediscovery of light. *Journal of Philosophy* 93 (5): 211–228.

Clark, A. 2013. Whatever next? Predictive brains, situated agents, and the future of cognitive science. *Behavioral and Brain Sciences* 36 (3): 181–204.

Clayton, P. 2006. Conceptual foundations of emergence theory. In *The Re-emergence of Emergence*, ed. P. Clayton and P. Davies, 1–31. Oxford: Oxford University Press.

Comas, D., F. Petit, and T. Preat. 2004. *Drosophila* long-term memory formation involves regulation of cathepsin activity. *Nature* 430 (6998): 460–463.

Cong, P., X. Ma, X. Hou, G. D. Edgecombe, and N. J. Strausfeld. 2014. Brain structure resolves the segmental affinity of anomalocaridid appendages. *Nature* 513:538–542.

Courtiol, E., and D. A. Wilson. 2014. Thalamic olfaction: Characterizing odor processing in the mediodorsal thalamus of the rat. *Journal of Neurophysiology* 111 (6): 1274–1285.

Craig, A. D. 2003a. A new view of pain as a homeostatic emotion. *Trends in Neurosciences* 26 (6): 303–307.

Craig, A. D. 2003b. Pain mechanisms: Labeled lines versus convergence in central processing. *Annual Review of Neuroscience* 26 (1): 1–30.

Craig, A. D. 2010. The sentient self. *Brain Structure and Function* 214: 563–577.

Crancher, P., M. G. King, A. Bennett, and R. B. Montgomery. 1972. Conditioning of a free operant in *Octopus cyanus* Grayi. *Journal of the Experimental Analysis of Behavior* 17 (3): 359–362.

Craver, C. 2007. Constitutive explanatory relevance. *Journal of Philosophical Research* 32:3–20.

Crick, F. 1995. *Astonishing Hypothesis: The Scientific Search for the Soul*. New York: Simon and Schuster.

Crick, F., and C. Koch. 2003. A framework for consciousness. *Nature Neuroscience* 6:119–126.

Cruse, H., and M. Schilling. 2015. Mental states as emergent properties: From walking to consciousness. In *Open Mind: 9(C)*, ed. T. Metzinger and J. Windt. Frankfurt am Main: MIND Group.

Cruse, H., and M. Schilling. 2016. No proof for subjective experience in insects. *Animal Sentience: An Interdisciplinary Journal on Animal Feeling* 1 (9): 13.

Damasio, A. R. 2000. *The Feeling of What Happens: Body and Emotion in the Making of Consciousness*. New York: Random House.

Damasio, A. 2010. *Self Comes to Mind: Constructing the Conscious Brain*. New York: Vintage.

Damasio, A., H. Damasio, and D. Tranel. 2012. Persistence of feelings and sentience after bilateral damage of the insula. *Cerebral Cortex* 23: 833–846.

Damasio, A. R., T. J. Grabowski, A. Bechara, H. Damasio, L. L. Ponto, J. Parvizi, and R. D. Hichwa. 2000. Subcortical and cortical brain activity during the feeling of self-generated emotions. *Nature Neuroscience* 3 (10): 1049–1056.

Dardis, A. 2008. *Mental Causation: The Mind–Body Problem*. New York: Columbia University Press.

Darmaillacq, A. S., L. Dickel, and J. Mather. 2014. *Cephalopod Cognition*. Cambridge: Cambridge University Press.

Davis, S. M., A. L. Thomas, K. J. Nomie, L. Huang, and H. A. Dierick. 2014. Tailless and Atrophin control *Drosophila* aggression by regulating neuropeptide signalling in the pars intercerebralis. *Nature Communications* 5:3177.

Deacon, T. W. 2011. *Incomplete Nature: How Mind Emerged from Matter*. New York: W. W. Norton.

De Arriba, M. D. C., and A. M. Pombal. 2007. Afferent connections of the optic tectum in lampreys: An experimental study. *Brain, Behavior and Evolution* 69:37–68.

De Bivort, B. L., and B. Van Swinderen. 2016. Evidence for selective attention in the insect brain. *Current Opinion in Insect Science* 15:9–15.

Dehaene, S. 2014. *Consciousness and the Brain: Deciphering How the Brain Codes Our Thoughts*. New York: Viking Penguin.

Dehaene, S., and Naccache, L. 2001. Towards a cognitive neuroscience of consciousness: Basic evidence and a workspace framework. *Cognition* 79: 1–37.

Del Bene, F., C. Wyart, E. Robles, A. Tran, L. Looger, E. K. Scott, et al. 2010. Filtering of visual information in the tectum by an identified neural circuit. *Science* 330 (6004): 669–673.

Dennett, D. C. 1988. Quining qualia. In *Consciousness in Contemporary Science*, ed. A. J. Marcel and E. Bisiach, 42–77. Oxford: Clarendon Press.

Dennett, D. C. 1991. *Consciousness Explained*. Boston: Little, Brown.

Denton, D. 2006. *The Primordial Emotions: The Dawning of Consciousness*. Oxford: Oxford University Press.

Dicke, U., and G. Roth. 2009. Evolution of the visual system in amphibians. In *Encyclopedia of Neurosciences*, ed. M. D. Binder, N. Hirokawa, and U. Windhorst, 1455–1459. Berlin: Springer.

Dretske, F. 1995. *Naturalizing the Mind*. Cambridge, MA: MIT Press.

Dugas-Ford, J., J. J. Rowell, and C. W. Ragsdale. 2012. Cell-type homologies and the origins of the neocortex. *Proceedings of the National Academy of Sciences of the United States of America* 109 (42): 16974–16979.

Edelman, D. B., B. J. Baars, and A. K. Seth. 2005. Identifying hallmarks of consciousness in non-mammalian species. *Consciousness and Cognition* 14 (1): 169–187.

Edelman, D. B., and A. K. Seth. 2009. Animal consciousness: A synthetic approach. *Trends in Neurosciences* 32 (9): 476–484.

Edelman, G. M. 1989. *The Remembered Present: A Biological Theory of Consciousness*. New York: Basic Books.

Edelman, G. M. 1992. *Bright Air, Brilliant Fire: On the Matter of the Mind*. New York: Basic Books.

Edelman, G. M. 2003. Naturalizing consciousness: A theoretical framework. *Proceedings of the National Academy of Sciences of the United States of America* 100:5520–5524.

Edelman, G. M., J. A. Gally, and B. J. Baars. 2011. Biology of consciousness. *Frontiers in Psychology* 2:4.

Edelman, S., R. Moyal, and T. Fekete. 2016. To bee or not to bee? *Animal Sentience: An Interdisciplinary Journal on Animal Feeling* 1 (9): 14.

Elwood, R. W. 2016. Might insects experience pain? *Animal Sentience: An Interdisciplinary Journal on Animal Feeling* 1 (9): 18.

Elwood, R. W., and M. Appel. 2009. Pain experience in hermit crabs? *Animal Behaviour* 77 (5): 1243–1246.

Erwin, D. H., and J. W. Valentine. 2013. *The Cambrian Explosion*. Greenwood Village, CO: Roberts and Company.

Eshel, N., M. Bukwich, V. Rao, V. Hemmelder, J. Tian, and N. Uchida. 2015. Arithmetic and local circuitry underlying dopamine prediction errors. *Nature* 525 (7568): 243–246.

Fang-Yen, C., M. J. Alkema, and A. D. Samuel. 2015. Illuminating neural circuits and behaviour in *Caenorhabditis elegans* with optogenetics. *Philosophical Transactions of the Royal Society B: Biological Sciences* 370 (1677): 20140212.

Fauria, K., M. Colborn, and T. S. Collett. 2000. The binding of visual patterns in bumblebees. *Current Biology* 10 (15): 935–938.

Feigl, H. 1967. *The "Mental" and the "Physical."* Minneapolis: University of Minnesota Press.

Feinberg, T. E. 2000. The nested hierarchy of consciousness: A neurobiological solution to the problem of mental unity. *Neurocase* 6:75–81.

Feinberg, T. E. 2001. Why the mind is not a radically emergent feature of the brain. *Journal of Consciousness Studies* 8 (9–10): 123–145.

Feinberg, T. E. 2009. *From Axons to Identity: Neurological Explorations of the Nature of the Self*. New York: W. W. Norton.

Feinberg, T. E. 2011. The nested neural hierarchy and the self. *Consciousness and Cognition* 20:4–17.

Feinberg, T. E. 2012. Neuroontology, neurobiological naturalism, and consciousness: A challenge to scientific reduction and a solution. *Physics of Life Reviews* 9 (1): 13–34.

Feinberg, T. E., and J. Mallatt. 2013. The evolutionary and genetic origins of consciousness in the Cambrian Period over 500 million years ago. *Frontiers in Psychology* 4.

Feinberg, T. E., and J. Mallatt. 2016a. *The Ancient Origins of Consciousness: How the Brain Created Experience*. Cambridge, MA: MIT Press.

Feinberg, T. E., and J. Mallatt. 2016b. The nature of primary consciousness: A new synthesis. *Consciousness and Cognition* 43:113–127.

Feinberg, T. E., and J. Mallatt. 2016c. The evolutionary origins of consciousness. In *Biophysics of Consciousness: A Foundational Approach*, ed. R. R. Poznanski, J. Tuszynski, and T. E. Feinberg, 47–86. London: World Scientific.

Felsenberg, J., O. Barnstedt, P. Cognigni, S. Lin, and S. Waddell. 2017. Re-evaluation of learned information in *Drosophila*. *Nature* 544 (7649): 240–244.

Fink, S. B. 2016. A deeper look at the "neural correlate of consciousness." *Frontiers in Psychology* 7.

Gans, C., and R. G. Northcutt. 1983. Neural crest and the origin of vertebrates: A new head. *Science* 220 (4594): 268–273.

Gennaro, R. J. 2011. *The Consciousness Paradox: Consciousness, Concepts, and Higher-Order Thoughts*. Cambridge, MA: MIT Press.

Gershman, S. J., E. J. Horvitz, and J. B. Tenenbaum. 2015. Computational rationality: A converging paradigm for intelligence in brains, minds, and machines. *Science* 349 (6245): 273–278.

Ginsburg, S., and E. Jablonka. 2010. The evolution of associative learning: A factor in the Cambrian explosion. *Journal of Theoretical Biology* 266 (1): 11–20.

Gillette, R., and J. W. Brown. 2015. The sea slug, *Pleurobranchaea californica*: A signpost species in the evolution of complex nervous systems and behavior. *Integrative and Comparative Biology* 55 (6): 1058–1069.

Giordano, J. 2005. The neurobiology of nociceptive and antinociceptive systems. *Pain Physician* 8 (3): 277–290.

Gizowski, C., and C. W. Bourque. 2017. Neurons that drive and quench thirst. *Science* 357 (6356): 1092–1093.

Globus, G. G. 1973. Unexpected symmetries in the "world knot." *Science* 180:1129–1136.

Godfrey-Smith, P. 2016a. *Other Minds: The Octopus, the Sea, and the Deep Origins of Consciousness*. London: Macmillan.

Godfrey-Smith, P. 2016b. Animal evolution and the origins of experience. In *How Biology Shapes Philosophy: New Foundations for Naturalism*, ed. D. L. Smith, 23–50. Cambridge: Cambridge University Press.

Graham, B. J., and D. P. Northmore. 2007. A spiking neural network model of midbrain visuomotor mechanisms that avoids objects by estimating size and distance monocularly. *Neurocomputing* 70 (10): 1983–1987.

Graindorge, N., C. Alves, A. S. Darmaillacq, R. Chichery, L. Dickel, and C. Bellanger. 2006. Effects of dorsal and ventral vertical lobe electrolytic lesions on spatial learning and locomotor activity in *Sepia officinalis*. *Behavioral Neuroscience* 120 (5): 1151.

Grewe, B. F., J. Gründemann, L. J. Kitch, J. A. Lecoq, J. G. Parker, J. D. Marshall, et al. 2017. Neural ensemble dynamics underlying a long-term associative memory. *Nature* 543 (7647): 670–675.

Grillner, S., J. Hellgren, A. Menard, K. Saitoh, and M. A. Wikström. 2005. Mechanisms for selection of basic motor programs—roles for the striatum and pallidum. *Trends in Neurosciences* 28 (7): 364–370.

Gruberg, E., E. Dudkin, Y. Wang, G. Marín, C. Salas, E. Sentis, et al. 2006. Influencing and interpreting visual input: The role of a visual feedback system. *Journal of Neuroscience* 26 (41): 10368–10371.

Guirado, S., and J. C. Davila. 2009. Evolution of the optic tectum in amniotes. In *Encyclopedia of Neurosciences*, ed. M. D. Binder, N. Hirokawa, and U. Windhorst, 1375–1380. Berlin: Springer.

Gutfreund, Y. 2012. Stimulus-specific adaptation, habituation, and change detection in the gaze control system. *Biological Cybernetics* 106 (11–12): 657–668.

Gutnick, T., R. A. Byrne, B. Hochner, and M. Kuba. 2011. *Octopus vulgaris* uses visual information to determine the location of its arm. *Current Biology* 21 (6): 460–462.

Hall, B. K. 2008. *The Neural Crest and Neural Crest Cells in Vertebrate Development and Evolution*. 2nd ed. New York: Springer Science & Business Media.

Hall, J. 2011. *Guyton and Hall Textbook of Medical Physiology*. 12th ed. Philadelphia: Saunders.

Haralson, J. V., C. I. Groff, and S. J. Haralson. 1975. Classical conditioning in the sea anemone, *Cribrina xanthogrammica*. *Physiology and Behavior* 15 (4): 455–460.

Hardisty, M. W. 1979. *Biology of the Cyclostomes*. London: Chapman & Hall.

Harnad, S. 2016. Animal sentience: The other-minds problem. *Animal Sentience: An Interdisciplinary Journal on Animal Feeling* 1 (1): 1.

Heil, J., and A. Mele, eds. 1993. *Mental Causation*. Oxford: Clarendon Press.

Herberholz, J., and G. D. Marquart. 2012. Decision making and behavioral choice during predator avoidance. *Frontiers in Neuroscience* 6.

Hills, T. T. 2006. Animal foraging and the evolution of goal-directed cognition. *Cognitive Science* 30 (1): 3–41.

Hochner, B. 2012. An embodied view of octopus neurobiology. *Current Biology* 22 (20): R887–R892.

Hochner, B. 2013. How nervous systems evolve in relation to their embodiment: What we can learn from octopuses and other molluscs. *Brain, Behavior and Evolution* 82 (1): 19–30.

Hochner, B., T. Shomrat, and G. Fiorito. 2006. The octopus: A model for a comparative analysis of the evolution of learning and memory mechanisms. *Biological Bulletin* 210 (3): 308–317.

Hodos, W., and A. B. Butler. 1997. Evolution of sensory pathways in vertebrates. *Brain, Behavior and Evolution* 50 (4): 189–197.

Hohwy, J. 2007. The search for neural correlates of consciousness. *Philosophy Compass* 2 (3): 461–474.

Homberg, U. 2008. Evolution of the central complex in the arthropod brain with respect to the visual system. *Arthropod Structure and Development* 37 (5): 347–362.

Hu, H. 2016. Reward and aversion. *Annual Review of Neuroscience* 39:297–324.

Huber, R., J. B. Panksepp, T. Nathaniel, A. Alcaro, and J. Panksepp. 2011. Drug-sensitive reward in crayfish: An invertebrate model system for the study of seeking, reward, addiction, and withdrawal. *Neuroscience and Biobehavioral Reviews* 35 (9): 1847–1853.

Hume, J. B., C. E. Adams, B. Mable, and C. W. Bean. 2013. Sneak male mating tactics between lampreys (Petromyzontiformes) exhibiting alternative life-history strategies. *Journal of Fish Biology* 82 (3): 1093–1100.

Huston, J. P., M. A. Silva, B. Topic, and C. P. Müller. 2013. What's conditioned in conditioned place preference? *Trends in Pharmacological Sciences* 34 (3): 162–166.

Jabr, F. 2012. Know your neurons: How to classify different types of neurons in the brain's forest. http://blogs.scientificamerican.com/brainwaves/2012/05/16/know-your-neurons-classifying-the-many-types-of-cells-in-the-neuron-forest.

Jackson, F. 1982. Epiphenomenal qualia. *Philosophical Quarterly* 32: 127–136.

James, W. 1904. Does consciousness exist? *Journal of Philosophy, Psychology, and Scientific Methods* 1 (18): 477–491.

Jarvis, E. D., J. Yu, M. V. Rivas, H. Horita, G. Feenders, O. Whitney, et al. 2013. Global view of the functional molecular organization of the avian cerebrum: Mirror images and functional columns. *Journal of Comparative Neurology* 521 (16): 3614–3665.

Jing, J., E. C. Cropper, I. Hurwitz, and K. R. Weiss. 2004. The construction of movement with behavior-specific and behavior-independent modules. *Journal of Neuroscience* 24 (28): 6315–6325.

Jonkisz, J. 2015. Consciousness: Individuated information in action. *Frontiers in Psychology* 6.

Kaas, A. L., M. C. Stoeckel, and R. Goebel. 2008. The neural bases of haptic working memory. In *Human Haptic Perception: Basics and Applications*, ed. M. Grunwald, 113–129. Boston: Birkhauser.

Kaas, J. H. 1997. Topographic maps are fundamental to sensory processing. *Brain Research Bulletin* 44 (2): 107–112.

Kandel, E. R., J. H. Schwartz, T. M. Jessell, S. A. Siegelbaum, and A. J. Hudspeth. 2012. *Principles of Neural Science*. 5th ed. New York: McGraw-Hill.

Kardamakis, A. A., J. Pérez-Fernández, and S. Grillner. 2016. Spatiotemporal interplay between multisensory excitation and recruited inhibition in the lamprey optic tectum. *eLife* 5:e16472.

Kardong, K. 2012. *Vertebrates: Comparative Anatomy, Function, Evolution*. 6th ed. Dubuque, IA: McGraw-Hill Higher Education.

Karten, H. J. 2013. Neocortical evolution: Neuronal circuits arise independently of lamination. *Current Biology* 23 (1): R12–R15.

Kawai, N., R. Kono, and S. Sugimoto. 2004. Avoidance learning in the crayfish (*Procambarus clarkii*) depends on the predatory imminence of the unconditioned stimulus: A behavior systems approach to learning in invertebrates. *Behavioural Brain Research* 150 (1): 229–237.

Keller, A. 2014. The evolutionary function of conscious information processing is revealed by its task-dependency in the olfactory system. *Frontiers in Psychology* 5.

Key, B. 2014. Fish do not feel pain and its implications for understanding phenomenal consciousness. *Biology and Philosophy* 30:149–165.

Key, B. 2016. Phenomenal consciousness in insects? A possible way forward. *Animal Sentience: An Interdisciplinary Journal on Animal Feeling* 1 (9): 17.

Khodagholy, D., J. N. Gelinas, and G. Buzsáki. 2017. Learning-enhanced coupling between ripple oscillations in association cortices and hippocampus. *Science* 358 (6361): 369–372.

Kim, J. 1992. "Downward causation" in emergentism and nonreductive physicalism. In *Emergence or Reduction? Essays on the Prospects of Nonreductive Physicalism*, ed. A. Beckermann, H. Flohr, and J. Kim, 119–138. New York: Walter de Gruyter.

Kim, J. 1998. *Mind in a Physical World: An Essay on the Mind–Body Problem and Mental Causation*. Cambridge, MA: MIT Press.

Kim, J. 2006. Being realistic about emergence. In *The Re-emergence of Emergence*, ed. P. Clayton and P. Davies, 190–202. Oxford: Oxford University Press.

Kim, S. S., H. Rouault, S. Druckmann, and V. Jayaraman. 2017. Ring attractor dynamics in the *Drosophila* central brain. *Science* 356 (6340): 849–853.

Kirk, R. 1994. *Raw Feeling*. Cambridge, MA: MIT Press.

Kisch, J., and J. Erber. 1999. Operant conditioning of antennal movements in the honey bee. *Behavioural Brain Research* 99 (1): 93–102.

Klein, C., and A. B. Barron. 2016a. Insects have the capacity for subjective experience. *Animal Sentience: An Interdisciplinary Journal on Animal Feeling* 1 (9): 1.

Klein, C., and A. B. Barron. 2016b. Insect consciousness: Commitments, conflicts and consequences. *Animal Sentience: An Interdisciplinary Journal on Animal Feeling* 1 (9): 21.

Klein, C., and A. B. Barron. 2016c. Reply to Adamo, Key et al., and Schilling and Cruse: Crawling around the hard problem of consciousness. *Proceedings of the National Academy of Sciences of the United States of America* 113 (27): E3814–E3815.

Knudsen, E. I. 2011. Control from below: The role of a midbrain network in spatial attention. *European Journal of Neuroscience* 33 (11): 1961–1972.

Koch, C. 2004. *The Quest for Consciousness: A Neurobiological Approach.* Englewood, CO: Roberts.

Koch, C., M. Massimini, M. Boly, and G. Tononi. 2016. Neural correlates of consciousness: Progress and problems. *Nature Reviews Neuroscience* 17 (5): 307–321.

Kohl, J. V. 2013. Nutrient-dependent/pheromone-controlled adaptive evolution: A model. *Socioaffective Neuroscience and Psychology* 3:20553.

Krauzlis, R. J., L. P. Lovejoy, and A. Zénon. 2013. Superior colliculus and visual spatial attention. *Annual Review of Neuroscience* 36:165–182.

Krebber, M., J. Harwood, B. Spitzer, J. Keil, and D. Senkowski. 2015. Visuotactile motion congruence enhances gamma-band activity in visual and somatosensory cortices. *NeuroImage* 117:160–169.

Kröger, B., J. Vinther, and D. Fuchs. 2011. Cephalopod origin and evolution: A congruent picture emerging from fossils, development, and molecules. *BioEssays* 33 (8): 602–613.

Kuba, M., T. Gutnick, and B. Hochner. 2012. Meeting an alien: Behavioral experiments on the octopus. In *Frontiers in Behavioral Neuroscience Conference Abstract: Tenth International Congress of Neuroethology*, vol. 436. doi:10.3389/conf.fnbeh.

Lacalli, T. C. 2008. Basic features of the ancestral chordate brain: A protochordate perspective. *Brain Research Bulletin* 75:319–323.

Lacalli, T. C. 2013. Looking into eye evolution: Amphioxus points the way. *Pigment Cell and Melanoma Research* 26:162–164.

Lacalli, T. C. 2015. The origin of vertebrate neural organization. In *Structure and Evolution of Invertebrate Nervous Systems,* ed. A. Schmidt-Rhaesa, S. Harszch, and G. Purschke. Oxford: Oxford University Press.

Lacalli, T. C. 2018. Amphioxus neurocircuits, enhanced arousal, and the origin of vertebrate consciousness. *Consciousness and Cognition* 62:127–134.

Lamb, T. D. 2013. Evolution of phototransduction, vertebrate photoreceptors, and retina. *Progress in Retinal and Eye Research* 36:52–119.

Lamme, V. A. 2006. Towards a true neural stance on consciousness. *Trends in Cognitive Sciences* 10 (11): 494–501.

Lamme, V. A., and P. R. Roelfsema. 2000. The distinct modes of vision offered by feedforward and recurrent processing. *Trends in Neurosciences* 23:571–579.

Lau, H., and D. Rosenthal. 2011. Empirical support for higher-order theories of conscious awareness. *Trends in Cognitive Sciences* 15 (8): 365–373.

LeDoux, J. 2012. Rethinking the emotional brain. *Neuron* 73 (4): 653–676.

LeDoux, J. E., and R. Brown. 2017. A higher-order theory of emotional consciousness. *Proceedings of the National Academy of Sciences of the United States of America* 114 (10): E2016–E2025.

Lee, S. H., and Y. Dan. 2012. Neuromodulation of brain states. *Neuron* 76 (1): 209–222.

Levine, J. 1983. Materialism and qualia: The explanatory gap. *Pacific Philosophical Quarterly* 64 (4): 354–361.

Llinás, R. R. 2002. *I of the Vortex: From Neurons to Self.* Cambridge, MA: MIT Press.

Lockwood, M. 1993. The grain problem. In *Objections to Physicalism,* ed. H. M. Robinson. Oxford: Oxford University Press.

Loy, I., V. Fernández, and F. Acebes. 2006. Conditioning of tentacle lowering in the snail (*Helix aspersa*): Acquisition, latent inhibition, overshadowing, second-order conditioning, and sensory preconditioning. *Learning and Behavior* 34 (3): 305–314.

Ma, X., X. Hou, G. D. Edgecombe, and N. J. Strausfeld. 2012. Complex brain and optic lobes in an early Cambrian arthropod. *Nature* 490 (7419): 258–261.

Mallatt, J., and T. E. Feinberg. 2016. Insect consciousness: Fine-tuning the hypothesis. *Animal Sentience: An Interdisciplinary Journal on Animal Feeling* 1 (9): 10.

Mallatt, J., and T. E. Feinberg. 2017. Consciousness is not inherent in but emergent from life. *Animal Sentience: An Interdisciplinary Journal on Animal Feeling* 11 (15): 1.

Mángano, M. G., and L. A. Buatois. 2014. Decoupling of body-plan diversification and ecological structuring during the Ediacaran–Cambrian transition: Evolutionary and geobiological feedbacks. In *Proceedings of the Royal Society B: Biological Sciences* 281 (1780): 20140038.

Manger, P. R. 2009. Evolution of the reticular formation. In *Encyclopedia of Neurosciences*, ed. M. D. Binder, N. Hirokawa, and U. Windhorst, 1413–1416. Berlin: Springer.

Marchetti, G. 2014. Attention and working memory: Two basic mechanisms for constructing temporal experiences. *Frontiers in Psychology* 5:880.

Marin, O., A. González, and W. J. Smeets. 1997. Anatomical substrate of amphibian basal ganglia involvement in visuomotor behaviour. *European Journal of Neuroscience* 9 (10): 2100–2109.

Marzluff, J. M., J. Walls, H. N. Cornell, J. C. Withey, and D. P. Craig. 2010. Lasting recognition of threatening people by wild American crows. *Animal Behaviour* 79 (3): 699–707.

Mather, J. 2012. Cephalopod intelligence. In *The Oxford Handbook of Comparative Evolutionary Psychology*, ed. J. Vonk and T. K. Shackelford, 118–128. Oxford: Oxford University Press.

Mather, J. A., and R. C. Anderson. 1999. Exploration, play and habituation in octopuses (*Octopus dofleini*). *Journal of Comparative Psychology* 113 (3): 333.

Mather, J. A., and C. Carere. 2016. Cephalopods are best candidates for invertebrate consciousness. *Animal Sentience: An Interdisciplinary Journal on Animal Feeling* 1 (9): 2.

Mather, J. A., and M. J. Kuba. 2013. The cephalopod specialties: Complex nervous system, learning, and cognition 1. *Canadian Journal of Zoology* 91 (6): 431–449.

Mayer, G. 2006. Structure and development of onychophoran eyes: What is the ancestral visual organ in arthropods? *Arthropod Structure and Development* 35 (4): 231–245.

Mayr, E. 1982. *The Growth of Biological Thought: Diversity, Evolution, and Inheritance*. Cambridge, MA: Harvard University Press.

Mayr, E. 2004. *What Makes Biology Unique? Considerations on the Autonomy of a Scientific Discipline*. Cambridge: Cambridge University Press.

McGinn, C. 1991. Consciousness and content. In *Mind and Common Sense: Philosophical Essays on Commonsense Psychology*, ed. R. J. Bogdan, 71–92. Cambridge: Cambridge University Press.

McHaffie, J. G., T. R. Stanford, B. E. Stein, V. Coizet, and P. Redgrave. 2005. Subcortical loops through the basal ganglia. *Trends in Neurosciences* 28 (8): 401–407.

Meehl, P. 1966. The compleat autocerebroscopist: A thought experiment on Professor Feigl's mind/body identity thesis. In *Mind, Matter, and Method*, ed. P. K. Feyerabend and G. Maxwell, 103–180. Minneapolis: University of Minnesota Press.

Meek, J. 1981. A golgi-electron microscopic study of goldfish optic tectum. I. Description of afferents, cell types, and synapses. *Journal of Comparative Neurology* 199 (2): 149–173.

Merker, B. 2007. Consciousness without a cerebral cortex: A challenge for neuroscience and medicine. *Behavioral and Brain Sciences* 30 (01): 63–81.

Merker, B. H. 2016. Insects join the consciousness fray. *Animal Sentience: An Interdisciplinary Journal on Animal Feeling* 1 (9): 4.

Metzinger, T. 2004. *Being No One: The Self-Model Theory of Subjectivity*. Cambridge, MA: MIT Press.

Milkowski, M. 2013. *Explaining the Computational Mind*. Cambridge: MIT Press.

Min, B. K. 2010. A thalamic reticular networking model of consciousness. *Theoretical Biology and Medical Modelling* 7 (10): 1–18.

Montgomery, S. 2015. *The Soul of an Octopus: A Surprising Exploration into the Wonder of Consciousness*. New York: Simon and Schuster.

Mormann, F., and C. Koch. 2007. Neural correlates of consciousness. http://www.scholarpedia.org/article/Neural_correlates_of_consciousness.

Morowitz, H. J. 2004. *The Emergence of Everything: How the World Became Complex*. New York: Oxford University Press.

Morsella, E. 2005. The function of phenomenal states: Supramodular interaction theory. *Psychological Review* 112 (4): 1000.

Morsella, E., C. A. Godwin, T. K. Jantz, S. C. Krieger, and A. Gazzaley. 2016. Homing in on consciousness in the nervous system: An action-based synthesis. *Behavioral and Brain Sciences* 39: e168.

Morsella, E., and Z. Reyes. 2016. The difference between conscious and unconscious brain circuits. *Animal Sentience: An Interdisciplinary Journal on Animal Feeling* 1 (11): 10.

Nagel, T. 1974. What is it like to be a bat? *Philosophical Review* 83 (4): 435–450.

Namburi, P., R. Al-Hasani, G. G. Calhoon, M. R. Bruchas, and K. M. Tye. 2016. Architectural representation of valence in the limbic system. *Neuropsychopharmacology* 41 (7): 1697–1715.

Nevin, L. M., E. Robles, H. Baier, and E. K. Scott. 2010. Focusing on optic tectum circuitry through the lens of genetics. *BMC Biology* 8 (1): 126.

Newport, C., G. Wallis, Y. Reshitnyk, and U. E. Siebeck. 2016. Discrimination of human faces by archerfish (*Toxotes chatareus*). *Scientific Reports* 6:27523.

Nieuwenhuys, R., J. G. Veening, and P. Van Domburg. 1987. Cores and paracores: Some new chemoarchitectural entities in the mammalian neuraxis. *Acta Morphologica Neerlando-Scandinavica* 26:131.

Nonomura, K., S. H. Woo, R. B. Chang, A. Gillich, Z. Qiu, A. G. Francisco, et al. 2017. Piezo2 senses airway stretch and mediates lung inflation-induced apnoea. *Nature* 541 (7636): 176–181.

Northcutt, R. G. 2005. The new head revisited. *Journal of Experimental Zoology* 304B:274–297.

Northcutt, R. G., and H. Wicht. 1997. Afferent and efferent connections of the lateral and medial pallia of the silver lamprey. *Brain, Behavior and Evolution* 49 (1): 1–19.

Northmore, D. 2011. The optic tectum. In *The Encyclopedia of Fish Physiology: From Genome to Environment*, ed. A. Farrell, 131–142. San Diego, CA: Academic Press.

Nutman, A. P., V. C. Bennett, C. R. Friend, M. J. Van Kranendonk, and A. R. Chivas. 2016. Rapid emergence of life shown by discovery of 3,700-million-year-old microbial structures. *Nature* 537 (7621): 535–538.

Oberheim, N. A., S. A. Goldman, and M. Nedergaard . 2012. Heterogeneity of astrocytic form and function. In *Astrocytes: Methods and Protocols, Methods in Molecular Biology*, vol. 814, ed. Richard Milner, 23–45. New York: Humana Press.

O'Connell, L. A., and H. A. Hofmann. 2011. The vertebrate mesolimbic reward system and social behavior network: A comparative synthesis. *Journal of Comparative Neurology* 519 (18): 3599–3639.

Ortega-Hernández, J. 2015. Lobopodians. *Current Biology* 25 (19): R873–R875.

Packard, A., and J. T. Delafield-Butt. 2014. Feelings as agents of selection: Putting Charles Darwin back into (extended neo-)Darwinism. *Biological Journal of the Linnean Society* 112 (2): 332–353.

Pain, S. P. 2009. Signs of anger: Representation of agonistic behaviour in invertebrate cognition. *Biosemiotics* 2 (2): 181–191.

Panksepp, J. 1998. *Affective Neuroscience: The Foundations of Human and Animal Emotions*. New York: Oxford University Press.

Panksepp, J. 2005. Affective consciousness: Core emotional feelings in animals and humans. *Consciousness and Cognition* 14 (1): 30–80.

Panksepp, J. 2015. Affective preclinical modeling of psychiatric disorders: Taking imbalanced primal emotional feelings of animals seriously in our search for novel antidepressants. *Dialogues in Clinical Neuroscience* 17 (4): 363.

Panksepp, J. 2016. The cross-mammalian neurophenomenology of primal emotional affects: From animal feelings to human therapeutics. *Journal of Comparative Neurology* 524 (8): 1624–1635.

Panksepp, J., L. Normansell, J. F. Cox, and S. M. Siviy. 1994. Effects of neonatal decortication on the social play of juvenile rats. *Physiology and Behavior* 56 (3): 429–443.

Papini, M. R., and M. E. Bitterman. 1991. Appetitive conditioning in *Octopus cyanea*. *Journal of Comparative Psychology* 105 (2): 107.

Parker, A. 2009. *In the Blink of an Eye: How Vision Sparked the Big Bang of Evolution*. New York: Basic Books.

Parvizi, J., and A. R. Damasio. 2003. Neuroanatomical correlates of brainstem coma. *Brain* 126 (7): 1524–1536.

Pattee, H. H. 1970. The problem of biological hierarchy. In *Towards a Theoretical Biology*, vol. 3, ed. C. H. Waddington, 117–136. Chicago: Aldine.

Paulk, A. C., Y. Zhou, P. Stratton, L. Liu, and B. van Swinderen. 2013. Multichannel brain recordings in behaving *Drosophila* reveal oscillatory activity and local coherence in response to sensory stimulation and circuit activation. *Journal of Neurophysiology* 110 (7): 1703–1721.

Pecoits, E., K. O. Konhauser, N. R. Aubet, L. M. Heaman, G. Veroslavsky, R. A. Stern, and M. K. Gingras. 2012. Bilaterian burrows and grazing behavior at >585 million years ago. *Science* 336 (6089): 1693–1696.

Peirs, C., and R. P. Seal. 2016. Neural circuits for pain: Recent advances and current views. *Science* 354 (6312): 578–584.

Perry, C. J., L. Baciadonna, and L. Chittka. 2016. Unexpected rewards induce dopamine-dependent positive emotion—like state changes in bumblebees. *Science* 353 (6307): 1529–1531.

Perry, C. J., and A. B. Barron. 2013. Neural mechanisms of reward in insects. *Annual Review of Entomology* 58:543–562.

Perry, C. J., A. B. Barron, and K. Cheng. 2013. Invertebrate learning and cognition: Relating phenomena to neural substrate. *Wiley Interdisciplinary Reviews: Cognitive Science* 4 (5): 561–582.

Piccinini, G. 2015. *Physical Computation: A Mechanistic Account*. Oxford: Oxford University Press.

Piccinini, G., and C. Craver. 2011. Integrating psychology and neuroscience: Functional analyses as mechanism sketches. *Synthese* 183 (3): 283–311.

Pigliucci, M. 2013. What hard problem? *Philosophy Now* 99. https://philpapers.org/archive/PIGWHP.pdf.

Pirri, J. K., A. D. McPherson, J. L. Donnelly, M. M. Francis, and M. J. Alkema. 2009. A tyramine-gated chloride channel coordinates distinct motor programs of a *Caenorhabditis elegans* escape response. *Neuron* 62 (4): 526–538.

Plotnick, R. E., S. Q. Dornbos, and J. Chen. 2010. Information landscapes and sensory ecology of the Cambrian radiation. *Paleobiology* 36:303–317.

Pombal, M. A., and L. Puelles. 1999. Prosomeric map of the lamprey forebrain based on calretinin immunocytochemistry, Nissl stain, and ancillary markers. *Journal of Comparative Neurology* 414 (3): 391–422.

Prados, J., B. Alvarez, F. Acebes, I. Loy, J. Sansa, and M. M. Moreno-Fernández. 2013. Blocking in rats, humans and snails using a within-subjects design. *Behavioural Processes* 100:23–31.

Preuss, S. J., C. A. Trivedi, C. M. vom Berg-Maurer, S. Ryu, and J. H. Bollmann. 2014. Classification of object size in retinotectal microcircuits. *Current Biology* 24 (20): 2376–2385.

Proske, U., and S. C. Gandevia. 2012. The proprioceptive senses: Their roles in signaling body shape, body position and movement, and muscle force. *Physiological Reviews* 92 (4): 1651–1697.

Randall, F. E., M. A. Whittington, and M. O. Cunningham. 2011. Fast oscillatory activity induced by kainate receptor activation in the rat basolateral amygdala in vitro. *European Journal of Neuroscience* 33 (5): 914–922.

Reggia, J. A. 2013. The rise of machine consciousness: Studying consciousness with computational models. *Neural Networks* 44:112–131.

Revonsuo, A. 2006. *Inner Presence: Consciousness as a Biological Phenomenon*. Cambridge, MA: MIT Press.

Revonsuo, A. 2010. *Consciousness: The Science of Subjectivity*. Hove, UK: Psychology Press.

Ristau, C. A. 2016. Beginnings: Physics, sentience and LUCA. *Animal Sentience: An Interdisciplinary Journal on Animal Feeling* 1 (11): 4.

Robertson, B., K. Saitoh, A. Ménard, and S. Grillner. 2006. Afferents of the lamprey optic tectum with special reference to the GABA input: Combined tracing and immunohistochemical study. *Journal of Comparative Neurology* 499 (1): 106–119.

Rodríguez-Moldes, I., P. Molist, F. Adrio, M. A. Pombal, S. E. P. Yáñez, M. Mandado, et al. 2002. Organization of cholinergic systems in the brain of different fish groups: A comparative analysis. *Brain Research Bulletin* 57 (3): 331–334.

Rolls, E. T. 2014. Emotion and decision-making explained: Précis; Synopsis of book published by Oxford University Press, 2014. *Cortex* 59:185–193.

Ruppert, E., R. Fox, and R. Barnes. 2004. *Invertebrate Zoology: A Functional Evolutionary Approach.* 7th ed. Belmont, CA: Thomson/Brooks/Cole.

Ryan, K., Z. Lu, and I. A. Meinertzhagen. 2016. The CNS connectome of a tadpole larva of *Ciona intestinalis* highlights sidedness in the brain of a chordate sibling. *eLife* 5:e16962.

Ryczko, D., S. Grätsch, F. Auclair, C. Dubé, S. Bergeron, M. H. Alpert, et al. 2013. Forebrain dopamine neurons project down to a brainstem region controlling locomotion. *Proceedings of the National Academy of Sciences of the United States of America* 110:E3235–E3242.

Saidel, W. M. 2009. Evolution of the optic tectum in anamniotes. In *Encyclopedia of Neurosciences,* ed. M. D. Binder, N. Hirokawa, and U. Windhorst, 1380–1387. Berlin: Springer.

Saitoh, K., A. Ménard, and S. Grillner. 2007. Tectal control of locomotion, steering, and eye movements in lamprey. *Journal of Neurophysiology* 97 (4): 3093–3108.

Salthe, S. N. 1985. *Evolving Hierarchical Systems: Their Structure and Representation.* New York: Columbia University Press.

Schiffbauer, J. D., J. W. Huntley, G. R. O'Neil, S. A. Darroch, M. Laflamme, and Y. Cai. 2016. The latest Ediacaran wormworld fauna: Setting the ecological stage for the Cambrian explosion. *GSA Today* 26 (11): 4–11.

Schlosser, G. 2014. Development and evolution of vertebrate cranial placodes. *Developmental Biology* 389:1.

Schluessel, V., and H. Bleckmann. 2005. Spatial memory and orientation strategies in the elasmobranch *Potamotrygon motoro. Journal of Comparative Physiology A: Neuroethology, Sensory, Neural, and Behavioral Physiology* 191 (8): 695–706.

Schopf, J. W., and A. B. Kudryavtsev. 2012. Biogenicity of Earth's earliest fossils: A resolution of the controversy. *Gondwana Research* 22 (3): 761–771.

Schuelert, N., and U. Dicke. 2005. Dynamic response properties of visual neurons and context-dependent surround effects on receptive fields in the tectum of the salamander *Plethodon shermani*. *Neuroscience* 134 (2): 617–632.

Schultz, W. 2015. Neuronal reward and decision signals: From theories to data. *Physiological Reviews* 95 (3): 853–951.

Schumacher, S., T. B. de Perera, and G. von der Emde. 2017. Electrosensory capture during multisensory discrimination of nearby objects in the weakly electric fish *Gnathonemus petersii*. *Scientific Reports* 7:43665.

Schumann, I., L. Hering, and G. Mayer. 2016. Immunolocalization of arthropsin in the onychophoran *Euperipatoides rowelli* (Peripatopsidae). *Frontiers in Neuroanatomy* 10:80.

Searle, J. 1992. *The Rediscovery of the Mind*. Cambridge, MA: MIT Press.

Searle, J. R. 1997. *The Mystery of Consciousness*. New York: New York Review of Books.

Searle, J. R. 2007. Dualism revisited. *Journal of Physiology* 101 (4): 169–178.

Searle, J. R. 2008. Neurobiological naturalism. In *The Blackwell Companion to Consciousness*, ed. M. Velmans and S. Schneider, 325–334. Hoboken, NJ: John Wiley & Sons.

Searle, J. R. 2016. Foreword: Addressing the hard problem of consciousness. In *Biophysics of Consciousness: A Foundational Approach*, ed. R. R. Poznanski, J. Tuszynski, and T. E. Feinberg. London: World Scientific.

Sellars, W. 1963. *Science, Perception and Reality*. London: Routledge and Kegan Paul.

Sellars, W. 1965. The identity approach to the mind–body problem. *Review of Metaphysics* 18(3): 430–451.

Selverston, A. I. 2010. Invertebrate central pattern generator circuits. *Philosophical Transactions of the Royal Society B: Biological Sciences* 365 (1551): 2329–2345.

Seth, A. K. 2009a. Explanatory correlates of consciousness: Theoretical and computational challenges. *Cognitive Computation* 1 (1): 50–63.

Seth, A. K. 2009b. Functions of consciousness. In *Elsevier Encyclopedia of Consciousness*, ed. W. P. Banks, 279–293. San Francisco: Elsevier.

Seth, A. K. 2013. Interoceptive inference, emotion, and the embodied self. *Trends in Cognitive Sciences* 17 (11): 565–573.

Seth, A. K., B. J. Baars, and D. B. Edelman. 2005. Criteria for consciousness in humans and other mammals. *Consciousness and Cognition* 14 (1): 119–139.

Shanahan, M. 2016. Consciousness as integrated perception, motivation, cognition, and action. *Animal Sentience: An Interdisciplinary Journal on Animal Feeling* 1 (9): 12.

Shepherd, G. M. 2007. Perspectives on olfactory processing, conscious perception, and orbitofrontal cortex. *Annals of the New York Academy of Sciences* 1121 (1): 87–101.

Sherrington, C. S. 1906. *The Integrative Action of the Nervous System.* Oxford: Oxford University Press.

Shigeno, S. 2017. Brain evolution as an information flow designer: The ground architecture for biological and artificial general intelligence. In *Brain Evolution by Design*, ed. S. Shigeno, Y. Murakami, and T. Nomura, 415–438. Tokyo: Springer Japan.

Shohat-Ophir, G., K. R. Kaun, R. Azanchi, H. Mohammed, and U. Heberlein. 2012. Sexual deprivation increases ethanol intake in *Drosophila. Science* 335 (6074): 1351–1355.

Shu, D. G., S. C. Morris, J. Han, Z. F. Zhang, K. Yasui, P. Janvier, et al. 2003. Head and backbone of the early Cambrian vertebrate *Haikouichthys. Nature* 421 (6922): 526–529.

Simon, H. A. 1962. The architecture of complexity. *Proceedings of the American Philosophical Society* 106 (6): 467–482.

Simon, H. A. 1973. The organization of complex systems. In *Hierarchy Theory: The Challenge of Complex Systems*, ed. H. H. Pattee, 1–27. New York: George Braziller.

Solms, M. 2013. The conscious id. *Neuro-psychoanalysis* 15 (1): 5–19.

Søvik, E., and A. B. Barron. 2013. Invertebrate models in addiction research. *Brain, Behavior and Evolution* 82 (3): 153–165.

Søvik, E., and C. Perry. 2016. The evolutionary history of consciousness. *Animal Sentience: An Interdisciplinary Journal on Animal Feeling* 1 (9): 19.

Søvik, E., C. J. Perry, and A. B. Barron. 2015. Insect Reward Systems: Comparing flies and bees. *Advances in Insect Physiology* 48:189–226.

Stein, B. E., and M. A. Meredith. 1993. *The Merging of the Senses*. Cambridge, MA: MIT Press.

Stephan, C., A. Wilkinson, and L. Huber. 2012. Have we met before? Pigeons recognise familiar human faces. *Avian Biology Research* 5 (2): 75–80.

Sterling, P., and S. Laughlin. 2015. *Principles of Neural Design*. Cambridge, MA: MIT Press.

Stevenson, P. A., and K. Schildberger. 2013. Mechanisms of experience dependent control of aggression in crickets. *Current Opinion in Neurobiology* 23 (3): 318–323.

Strausfeld, N. J. 2013. *Arthropod Brains: Evolution, Functional Elegance, and Historical Significance*. Cambridge, MA: Harvard University Press.

Suryanarayana, S. M., B. Robertson, P. Wallén, and S. Grillner. 2017. The lamprey pallium provides a blueprint of the mammalian layered cortex. *Current Biology* 27 (21): 3264–3277.

Swink, W. D. 2003. Host selection and lethality of attacks by sea lampreys (*Petromyzon marinus*) in laboratory studies. *Journal of Great Lakes Research* 29:307–319.

Tashiro, T., A. Ishida, M. Hori, M. Igisu, M. Koike, P. Méjean, et al. 2017. Early trace of life from 3.95 Ga sedimentary rocks in Labrador, Canada. *Nature* 549 (7673): 516–518.

Temizer, I., J. C. Donovan, H. Baier, and J. L. Semmelhack. 2015. A visual pathway for looming-evoked escape in larval zebrafish. *Current Biology* 25 (14): 1823–1834.

Thompson, E. 2007. *Mind in Life: Biology, Phenomenology, and the Sciences of Mind*. Cambridge, MA: Harvard University Press.

Tomina, Y., and M. Takahata. 2010. A behavioral analysis of force-controlled operant tasks in American lobster. *Physiology and Behavior* 101 (1): 108–116.

Tonoki, A., and R. L. Davis. 2015. Aging impairs protein-synthesis-dependent long-term memory in *Drosophila*. *Journal of Neuroscience* 35 (3): 1173–1180.

Tononi, G. 2011. The integrated information theory of consciousness: An updated account. *Archives Italiennes de Biologie* 150 (2–3): 56–90.

Tononi, G., and C. Koch . 2015. Consciousness: Here, there, and everywhere? *Philosophical Transactions of the Royal Society B: Biological Sciences* 370 (1668): 1–18.

Trestman, M. 2013. The Cambrian explosion and the origins of embodied cognition. *Biological Theory* 8 (1): 80–92.

Tsubouchi, A., T. Yano, T. K. Yokoyama, C. Murtin, H. Otsuna, and K. Ito. 2017. Topological and modality-specific representation of somatosensory information in the fly brain. *Science* 358 (6363): 615–623.

Tsuchiya, N., and J. van Boxtel. 2013. Introduction to research topic: Attention and consciousness in different senses. *Frontiers in Psychology* 4.

Tye, M. 2000. *Consciousness, Color, and Content*. Cambridge, MA: MIT Press.

Tye, M. 2016. Are insects sentient? *Animal Sentience: An Interdisciplinary Journal on Animal Feeling* 1 (9): 5.

Underwood, E. 2015. The brain's identity crisis. *Science* 349 (6248): 575–577.

Van Gulick, R. 2001. Reduction, emergence and other recent options on the mind–body problem: A philosophical overview. *Journal of Consciousness Studies* 8:1–34.

Van Swinderen, B., and R. Andretic. 2011. Dopamine in *Drosophila*: setting arousal thresholds in a miniature brain. *Proceedings of the Royal Society B: Biological Sciences*, rspb2010.2564.

Velmans, M., and S. Schneider, eds. 2008. *The Blackwell Companion to Consciousness*. Hoboken, NJ: John Wiley & Sons.

Verkhratsky, A., and V. Parpura. 2014. *Introduction to Neuroglia: Colloquium Series on Neuroglia in Biology and Medicine: From Physiology to Disease*. San Rafael, CA: Morgan & Claypool Life Sciences.

Verschure, P. F. 2016. Synthetic consciousness: The distributed adaptive control perspective. *Philosophical Transactions of the Royal Society B: Biological Sciences* 371 (1701): 20150448.

Vierck, C. J., B. L. Whitsel, O. V. Favorov, A. W. Brown, and M. Tommerdahl. 2013. Role of primary somatosensory cortex in the coding of pain. *Pain* 154 (3): 334–344.

Vopalensky, P., J. Pergner, M. Liegertova, E. Benito-Gutierrez, D. Arendt, and Z. Kozmik. 2012. Molecular analysis of the amphioxus frontal eye unravels the evolutionary origin of the retina and pigment cells of the vertebrate eye. *Proceedings of the National Academy of Sciences of the United States of America* 109 (38): 15383–15388.

Waddell, S. 2013. Reinforcement signalling in *Drosophila*: Dopamine does it all after all. *Current Opinion in Neurobiology* 23 (3): 324–329.

Walter, S., and H.-D. Heckmann, eds. 2003. *Physicalism and Mental Causation*. Exeter: Imprint Academic.

Wilczynski, W., and R. G. Northcutt. 1983. Connections of the bullfrog striatum: Afferent organization. *Journal of Comparative Neurology* 214 (3): 321–332.

Woodruff, M. L. 2017. Consciousness in teleosts: There is something it feels like to be a fish. *Animal Sentience: An Interdisciplinary Journal on Animal Feeling* 2 (13): 1.

Wullimann, M. F., and P. Vernier. 2009. Evolution of the brain in fishes. In *Encyclopedia of Neurosciences*, ed. M. D. Binder, N. Hirokawa, and U. Windhorst, 1318–1326. Berlin: Springer.

Yoshinaga, S., and K. Nakajima. 2017. A crossroad of neuronal diversity to build circuitry. *Science* 356 (6336): 376–377.

Zeisel, A., A. B. Muñoz-Manchado, S. Codeluppi, P. Lönnerberg, G. La Manno, A. Juréus, et al. 2015. Cell types in the mouse cortex and hippocampus revealed by single-cell RNA-seq. *Science* 347 (6226): 1138–1142.

Zeman, A. 2001. Consciousness. *Brain* 124:1263–1289.

Zieger, M. V., and V. B. Meyer-Rochow. 2008. Understanding the cephalic eyes of pulmonate gastropods: A review. *American Malacological Bulletin* 26 (1–2): 47–66.

Index

Page numbers followed by a "b," "f," or "g" indicate boxes, figures, or glossary entries, respectively.

MY COUNTRY